Large Viscous Boundary Layers for Noncharacteristic Nonlinear Hyperbolic Problems

MEMOIRS
of the
American Mathematical Society

Number 826

Large Viscous Boundary Layers for Noncharacteristic Nonlinear Hyperbolic Problems

Guy Métivier
Kevin Zumbrun

May 2005 • Volume 175 • Number 826 (second of 4 numbers) • ISSN 0065-9266

American Mathematical Society
Providence, Rhode Island

2000 *Mathematics Subject Classification.*
Primary 35L60; Secondary 35B35.

Library of Congress Cataloging-in-Publication Data

Métivier, Guy.

Large viscous boundary layers for noncharacteristic nonlinear hyperbolic problems / Guy Métivier, Kevin Zumbrun.

p. cm. — (Memoirs of the American Mathematical Society, ISSN 0065-9266 ; no. 826)

"Volume 175, number 826 (second of 4 numbers)."

Includes bibliographical references.

ISBN 0-8218-3649-8 (alk. paper)

1. Nonlinear boundary value problems. 2. Differential equations, Hyperbolic. 3. Differential equations, Nonlinear. I. Zumbrun, Kevin. II. Title. III. Series.

QA3.A57 no. 826
[QA379]
510 s—dc22
[515′.3535] 2005041981

Memoirs of the American Mathematical Society

This journal is devoted entirely to research in pure and applied mathematics.

Subscription information. The 2005 subscription begins with volume 173 and consists of six mailings, each containing one or more numbers. Subscription prices for 2005 are \$606 list, \$485 institutional member. A late charge of 10% of the subscription price will be imposed on orders received from nonmembers after January 1 of the subscription year. Subscribers outside the United States and India must pay a postage surcharge of \$31; subscribers in India must pay a postage surcharge of \$43. Expedited delivery to destinations in North America \$35; elsewhere \$130. Each number may be ordered separately; *please specify number* when ordering an individual number. For prices and titles of recently released numbers, see the New Publications sections of the *Notices of the American Mathematical Society.*

Back number information. For back issues see the *AMS Catalog of Publications.*

Subscriptions and orders should be addressed to the American Mathematical Society, P. O. Box 845904, Boston, MA 02284-5904, USA. *All orders must be accompanied by payment.* Other correspondence should be addressed to 201 Charles Street, Providence, RI 02904-2294, USA.

Memoirs of the American Mathematical Society is published bimonthly (each volume consisting usually of more than one number) by the American Mathematical Society at 201 Charles Street, Providence, RI 02904-2294, USA. Periodicals postage paid at Providence, RI. Postmaster: Send address changes to Memoirs, American Mathematical Society, 201 Charles Street, Providence, RI 02904-2294, USA.

This publication is indexed in *Science Citation Index*®, *SciSearch*®, *Research Alert*®, *CompuMath Citation Index*®, *Current Contents*®/*Physical, Chemical & Earth Sciences.*
Printed in the United States of America.

♾ The paper used in this book is acid-free and falls within the guidelines established to ensure permanence and durability.
Visit the AMS home page at `http://www.ams.org/`

10 9 8 7 6 5 4 3 2 1 10 09 08 07 06 05

Contents

Abstract

We study linear and nonlinear stability of large-amplitude multidimensional viscous boundary layers arising through the small viscosity perturbation of a hyperbolic initial–boundary value problem with noncharacteristic boundary. Our main result is to show that, provided there holds the necessary condition that all "frozen," planar boundary layers associated with the inner layer of the profile satisfy an appropriate Evans function condition, then the linearized equations about the full profile are well-posed in L^2, with sufficiently strong estimates on the solution and its derivatives as to yield a full nonlinear stability result and thereby nonlinear continuation/validation of the formal boundary layer expansion (alternatively, short-time existence for prepared initial data). The method of analysis is by symmetrizers and an appropriate extension of Kreiss' analysis of hyperbolic equations. Notable technical aspects include reduction to constant coefficients of the resolvent equation by an extension of the Gap Lemma of Evans function theory, clarification of the role of block structure in Kreiss-type estimates, and the use of conormal derivative estimates in the hyperbolic–parabolic setting.

Received by the editor March 26, 2002.

2000 *Mathematics Subject Classification.* 35L60 (35B35).

Key words and phrases. Boundary layers, noncharacteristic hyperbolic initial–boundary-value problems, Kreiss symmetrizers, Evans function.

G.M. thanks Indiana University for its hospitality during a visit in which this work was partially carried out.

K.Z. thanks the University of Rennes and E.N.S. Lyon for their hospitality during two visits in which this work was partially carried out. Thanks also to Mark Williams for his careful reading of early versions of the manuscript and many helpful suggestions/corrections. Research of K.Z. was partially supported under NSF grant number DMS-0070765.

Large Viscous Boundary Layers

1. Introduction

In this paper, we study the linear and nonlinear stability of viscous boundary layers which arise when one considers small viscosity parabolic perturbations of hyperbolic equations. For linear equations this problem is studied in [**BBB**], [**Ba-Ra**], [**Lio**]. The semilinear case is solved in [**Gu1**]. For quasilinear equations, a partial answer is given in [**Gr-Gu**] (see also [**Gi-Se**] for results in one space dimension). Indeed, the analysis in [**Gr-Gu**] has two parts. In the first part, approximate solutions are obtained using formal expansions in series of the viscosity ε. In the second part, the authors prove the stability of this approximate solution, proving that the exact solution is actually close to the approximate one, *using a smallness condition* (as in [**Gi-Se**]). By an example, they also show that some condition is needed. However, the smallness condition is not natural and does not allow large boundary layers.

The goal of this paper is to remove this smallness assumption, replacing it by an accurate assumption based on the analysis of an Evans function. Evans functions have been introduced in the study of the stability of planar viscous shock and boundary layers (see, e.g., [**GZ**], [**ZH**], [**ZS**], [**Z**], [**S**], [**Rou**], and references therein[1]). They play the role of the Lopatinski determinant for constant coefficient boundary value problems. When they vanish in the open left half plane, the problem is strongly unstable and when they do not vanish in the closed half space, the problem is expected to be strongly stable. Rescaling the variables, the results of [**ZH**], [**Z**] can be used to study the linear stability of boundary layers created by viscous perturbations of constant state solutions of hyperbolic equations on a half space providing some estimates of the corresponding Green's function. This indicates that assumptions on the Evans function should be the correct approach in the study of the stability of boundary layers. This has been proved to be correct in space dimension one [**Gr-Ro**] and the goal of this paper is to extend the analysis to multidimensional problems.

The one space dimensional analysis in [**Gr-Ro**] is based on integrations along characteristics for the hyperbolic equations and on pointwise estimates of the Green's function for the parabolic part, which are then combined to yield L^1 bounds on the Green's function for the linearized equations about the full boundary layer expansion. In multi-dimensions, both ingredients break down, due to more complicated geometry of characteristic surfaces. In particular, the known estimates of the parabolic Green's function [**Z**] consist of L^p bounds, $p \geq 2$, and do not include pointwise behavior. Moreover, it is known from study of the constant-coefficient

[1]For the origins of this method in the study of reaction–diffusion equations; see, e.g., [**E1**]-[**E4**], [**J**], [**AGJ**], [**PW**], [**K**].

case [**HoZ**] that the L^1 norm of the Green's function is *not* necessarily bounded in multi-dimensions, but in general may grow time-algebraically. This is a consequence of focusing and spreading in the underlying hyperbolic propagation, the effects of which are even more dramatic without parabolic regularization. Indeed, examples given by Rauch [**Ra1**] of L^p instability, $p \neq 2$, of smooth perturbations of constant states give reason to believe that L^2 is the only L^p norm in which we can expect that multi-dimensional hyperbolic problems be well-posed.

Thus, we are restricted in multi-dimensions by the hyperbolic (or "outer") part of the solution to seeking $L^2 \to L^2$ bounds, analogous to but (even in the constant-coefficient case) distinct from the L^1 Green's function bounds found in [**Gr-Ro**]. Moreover, we must obtain these bounds by a method suitable for the analysis of both hyperbolic boundary-value problems and their parabolic regularizations. To satisfy these requirements, we follow Kreiss' analysis of hyperbolic equations. Our basic estimate concerns the L^2 stability of the linearized equations, and is proved using symmetrizers and a suitable extension of Kreiss' analysis to parabolic-hyperbolic problems.

It can be seen by comparison with explicit representations of the resolvent in the planar case, carried out respectively in [**Ag**] and [**Z**], that this basic estimate is sharp *for both hyperbolic and parabolic parts of the equations.* Moreover, we significantly relax the structural assumptions under which the parabolic results of [**Z**] were obtained, just as the Kreiss analysis relaxed the assumptions necessary for the hyperbolic results of [**Ag**] .

Consider a first order quasilinear system

$$L(b,u,\partial)u := \partial_t u + \sum_{j=1}^{d} A_j(b,u)\partial_j u = F(b,u) \tag{1.1}$$

The equation holds on $\mathbb{R} \times \Omega$ where Ω is a regular bounded domain in $\mathbb{R}^d$. The unknown u is valued in $\mathbb{R}^N$ and $b = b(t,x)$ will be a given function which represents the variables (t,x) and the various possible source terms:

$$b(t,x) = (t,x,b_0(t,x)) \in \mathbb{R}^{1+d} \times \mathbb{R}^{M_0} = \mathbb{R}^M\,. \tag{1.2}$$

For example, one can think of (1.1) as a system of conservation laws with unknown b_0+u where b_0 is some background variable state and u a perturbation. The function b_0 can also appear as a forcing term in the right hand side. Since we will use change of coordinates, we also include the variables (t,x) in the coefficient b. The parameter b will vary in a domain $\mathcal{B} = [-T,T] \times \overline{\Omega} \times \mathcal{B}_0$ where $\mathcal{B}_0$ is a bounded open set in $\mathbb{R}^{M_0}$. The pair (b,u) will vary jointly within an open set $\mathcal{O} \subset \mathcal{B} \times \mathbb{R}^N$, having the form of a graph $\cup_{b\in\mathcal{B}}\mathcal{U}(b)$ over $\mathcal{B}$, where $\mathcal{U}(\cdot)$ is a continuous set-valued function from $\mathcal{B}$ to open sets in $\mathbb{R}^N$; in the simplest case, $\mathcal{U}(b) \equiv \mathcal{U}$ and $\mathcal{O} = \mathcal{B} \times \mathcal{U}$. (The latter suffices to discuss small amplitude boundary layers, i.e., u small, or, more generally, for boundary layers with small variation; see discussion below Definition 1.3.) We denote by $\mathcal{B}_\partial$ the set $\mathcal{B} = [-T,T] \times \partial\Omega \times \overline{\mathcal{B}}_0$, and $\mathcal{O}_\partial := \cup_{b\in\mathcal{B}_\partial}\mathcal{U}(b)$ the restriction of $\mathcal{O}$ to $\partial\Omega$. For $(b,u) = (t,x,b_0,u) \in \mathcal{O}_\partial$, we denote by

$$A_n(b,u) = \sum_{j=1}^{d} \nu_j(x)A_j(b,u) \tag{1.3}$$

the boundary matrix where $\nu(x) = (\nu_1(x),\ldots,\nu_d(x))$ is the inner unit normal vector to $\partial\Omega$ at x.

Next, we consider a parabolic viscous perturbation of (1.1)

$$L(b,u,\partial)u - \varepsilon \sum_{1\le j,k\le d} \partial_j\big(B_{j,k}(b,u)\partial_k u\big) = F(b,u)\,. \tag{1.4}$$

with Dirichlet boundary conditions:

$$u_{\,|\partial\Omega} = 0\,. \tag{1.5}$$

Note that nonhomogeneous boundary conditions reduce to homogeneous ones, changing u into $u_0 + v$, and adding u_0 to the parameters b_0. For the parabolic problem (1.5), (b,u) is allowed to range in a possibly larger open set $\mathcal{O}^*$ containing $\mathcal{O}$. For small amplitude boundary layers, we may take $\mathcal{O}^* = \mathcal{O} = \mathcal{B}\times\mathcal{U}$; however, in general we wish to allow the situation that the solution of (1.5), within the boundary layer near $\partial\Omega$, may take on values of (b,u) that lie outside the domain of well-posedness (i.e., hyperbolicity) of equation (1.1).

Assumption 1.1.

(H0) *The A_j and $B_{j,k}$ are $N\times N$ real matrices, C^∞ for (b,u) in $\mathcal{O}^*$; F is a smooth function from $\mathcal{O}^*$ to $\mathbb{R}^N$.*

(H1) *There is $c>0$ such that for all $(b,u)\in\mathcal{O}^*$ and all $\xi\in\mathbb{R}^d$ the eigenvalues of $\sum_{j,k=1}^d \xi_j\xi_k B_{j,k}(u)$ satisfy* $\operatorname{Re}\mu \ge c|\xi|^2$.

(H2) *For all $(b,u)\in\mathcal{O}$, the eigenvalues of $\sum\xi_j A_j(b,u)$ are real and semi-simple and have constant multiplicities for $(b,u)\in\mathcal{O} := \mathcal{B}\times\mathcal{U}$ and $\xi\in\mathbb{R}^d\setminus\{0\}$.*

(H3) *There is $c>0$ such that for all $(b,u)\in\mathcal{O}$ and $\xi\in\mathbb{R}^d$ the eigenvalues of $i\sum_{j=1}^d \xi_j A_j(u) + \sum_{j,k=1}^d \xi_j\xi_k B_{j,k}(u)$ satisfy* $\operatorname{Re}\mu \ge c|\xi|^2$.

(H4) *For all $(b,u)\in\mathcal{O}_\partial$, there holds* $\det A_n(b,u)\neq 0$.

The Assumption (H1) means that the perturbation

$$B(b,u,\partial) := \sum \partial_j\Big(B_{j,k}(b,u)\partial_k\,\cdot\Big)$$

is uniformly parabolic. (H2) means that L is hyperbolic, at least when the state (b,u) remains in the domain $\mathcal{O}$. The important Assumption (H4) means that the boundary $\partial\Omega$ is noncharacteristic for L. The Assumption (H3) is a compatibility condition between L and B. For example, when $B=\Delta_x$ is the Laplacian, (H1) is trivial and (H3) follows immediately from (H2). When (1.1) is a system of conservation laws which admits a strictly convex entropy $\eta(u)$, the system is symmetric hyperbolic. If in addition, $\operatorname{Re}\big(\eta''(u)\sum\xi_j\xi_k B_{j,k}(u)\big)$ is definite positive for all $\xi\neq 0$, then the assumptions (H1) and (H3) are satisfied.

The first step is to find the correct limiting boundary conditions for equation (1.1). These come from the study of a natural "inner layer" o.d.e equation, for $b\in\mathcal{B}_\partial$, (see [**Gi-Se**], [**Rou**],):

$$A_n(b,w)\frac{dw}{dz} - \frac{d}{dz}\Big(B_n(b,w)\frac{dw}{dz}\Big) = 0\,,\quad w(0)=0\,, \tag{1.6}$$

where $B_n(b,u) = \sum\nu_j(x)\nu_k(x)B_{j,k}(b,u)$. In (1.6) z stands for a fast variable in the direction ν. If $(y,x_n)\mapsto y + x_n\nu(y)$ with $y\in\partial\Omega$ and x_n small parametrizes a neighborhood of $\partial\Omega$, then z is a placeholder for x_n/ε. In what follows, what we call a solution of (1.6), is a solution on $[0,\infty[$ such that $(b,w(z))\in\mathcal{O}^*$ for all z.

One introduces the set

$$\mathcal{C} = \big\{(b,u) \in \mathcal{O}_\partial \ : \ (1.6) \text{ has a solution such that } u = \lim_{z\to+\infty} w(z)\big\} \tag{1.7}$$

Following [**Gi-Se**], [**Rou**], [**Gr-Gu**] the correct limit boundary conditions for (1.1) read

$$\forall x \in \partial\Omega\,, \quad (b(t,x), u(t,x)) \in \mathcal{C}\,. \tag{1.8}$$

Constants are solutions of (1.6) and Assumption (H4) implies that for all (b,u) there is a family of solutions $w_{b,u,v}(z)$, depending on parameters v, such that $w_{b,u,v}$ converges to u as z tends to infinity at an exponential rate (stable-central manifold theorem). That one of these solutions connects u to zero is a condition on (b,u) which defines $\mathcal{C}$. That the connection can be chosen smooth with respect to parameters is a transversality assumption.

To start the discussion, we first assume that for all $b \in \mathcal{B}_\partial$ a family of solutions of (1.6) is chosen, connecting 0 to a set of end states $\mathcal{C}_b \subset \mathcal{U}(b)$.

ASSUMPTION 1.2. *We are given a smooth manifold $\mathcal{C} \subset \mathcal{O}_\partial$ such that for all $b \in \mathcal{B}_\partial$, $\mathcal{C}_b := \{u \in \mathcal{U}(b) : (b,u) \in \mathcal{C}\} \neq \emptyset$, and a smooth function W from $\mathcal{C} \times [0,\infty[$ such that for all $(b,u) \in \mathcal{C}$, $w_{b,u} = W(b,u,\cdot)$ is a solution of* (1.6) *and $w_{b,u}(z)$ converges to u when z tends to $+\infty$, at an exponential rate, which can be chosen uniform on compact subsets of $\mathcal{C}$.*

Assumption 1.2 is the natural analog of assumption (H4), [**Z**], made in the planar shock theory.

The properties of $\mathcal{C}$ depend on the stability of $w_{b,u}$ solutions of ODE (1.6). Consider the linearized equation of (1.6) around $w_{b,u}$

$$\left\{\begin{aligned} &A_n(w)\frac{d\dot w}{dz} + (A_n'(w)\cdot \dot w)\frac{dw}{dz} - \frac{d}{dz}\Big(B_n(w)\frac{d\dot w}{dz}\Big) \\ &\qquad\qquad - \frac{d}{dz}\Big(\big(\dot w \cdot \nabla_u B_n'(w)\big)\frac{dw}{dz}\Big) = 0 \\ &\dot w(0) = 0\,. \end{aligned}\right. \tag{1.9}$$

These equations depend on the parameters $b \in \mathcal{B}_\partial$ and the function $w(z)$, with $w \in C^\infty([0,\infty[;\mathbb{R}^N)$. The unknown is $\dot w$.

DEFINITION 1.3. We say that the limiting boundary conditions (1.8) are transversal if:

i) For all $b \in \mathcal{B}_\partial$, $\mathcal{C}_b$ is a smooth manifold of dimension N_- equal to the number of negative eigenvalues of $A_n(b,u)$.

ii) For $(b,u) \in \mathcal{C}$, the tangent space of $\mathcal{C}_b$ at u is the set of $\dot u$ such that the linearized equation (1.9) with $w = w_{b,u}$ has a (unique) solution $\dot w$ such that $\dot u = \lim_{z\to\infty} \dot w(z)$.

In [**Gr-Gu**], it is proved that in the small amplitude case, i.e., for u in a suitably small neighborhood of 0, there is a unique manifold $\mathcal{C}$ and connection W having the properties above; moreover, the transversality condition is satisfied. They also prove that the boundary conditions (1.8) are maximally dissipative, when u is small and the parabolic term is the Laplace operator. In the large, we substitute for maximal dissipativity the more general uniform Kreiss-Lopatinski condition.

Consider a point $(\underline{b}, \underline{u}) \in \mathcal{C}$, $b = (\underline{t}, \underline{x}, \underline{b}_0)$. To derive the Kreiss-Lopatinski condition for hyperbolic problems, the idea is to approximate Ω near $\underline{x}$ by the half space $\{\nu(\underline{x})\cdot x > 0\}$, and to linearize the equation (1.1) around the constant solution u. This leads to a constant coefficient problem which is analyzed using a tangential Fourier-Laplace transform. We proceed in a similar way for (1.4). To $(b, u) \in \mathcal{C}$ we associate the profile $w(z) = w_{b,u}(z)$. The substitute for the constant state u is the "planar" boundary layer $\widetilde{w}(x) = w(\nu \cdot x/\varepsilon)$ that interpolates between 0 on the boundary $\nu \cdot x = 0$ and the inner state u . Next, we linearize the equation (1.4) around $\widetilde{w}$ to get the linear operator

$$\partial_t v + \sum_{j=1}^d A_j^\sharp \partial_j v - \varepsilon \sum_{j,k=1}^d \widetilde{B}_{j,k} \partial^2_{j,k} v + \frac{1}{\varepsilon} E^\sharp v \tag{1.10}$$

where

$$\begin{aligned} A_j^\sharp v &= \widetilde{A}_j v - \sum_{k=1}^d \nu_k (\partial_z \widetilde{w} \cdot \nabla_u \widetilde{B}_{k,j}) v - \sum_{k=1}^d \nu_k (v \cdot \nabla_u \widetilde{B}_{j,k}) \partial_z \widetilde{w} \\ E^\sharp v &= \sum_{k=1}^d \nu_k (v \cdot \nabla_u \widetilde{A}_k) \partial_z \widetilde{w} - \sum_{j,k=1}^d \nu_j \nu_k (v \cdot \nabla_u \widetilde{B}_{j,k}) \partial_z^2 \widetilde{w} \\ &\quad - \sum_{j,k=1}^d \nu_j \nu_k (\nabla_u^2 \widetilde{B}_{j,k}(v, \partial_z \widetilde{w})) \partial_z \widetilde{w} \end{aligned} \tag{1.11}$$

where $\widetilde{f}$ stands for the function f evaluated at $(b, \widetilde{w})$. In (1.10), the coefficients are functions of $\nu\cdot x$ only, and thus we can again perform a Fourier-Laplace transform in (t, y) where y are the variables in the tangent space $T_{\underline{x}}\Omega$. This leads us to introduce the following symbols, where $\eta \in T^*_{\underline{x}}\partial\Omega$ is a Fourier tangential frequency and $\tau - i\gamma$ a Fourier-Laplace time frequency:

$$M = (i\tau + \gamma)\mathrm{Id} + \sum_{j=1}^d i\eta_j A_j^\sharp + \sum_{1\le j,k\le d} \eta_j\eta_k \widetilde{B}_{j,k} + E^\sharp \tag{1.12}$$

$$A = \sum_{j=1}^d \nu_j A_j^\sharp - \sum_{1\le j,k\le d} i\eta_j \nu_k (\widetilde{B}_{j,k} + \widetilde{B}_{k,j}) \tag{1.13}$$

The symbols M and A are functions of $z \in [0, \infty[$ which depend on the parameters $(b, u) \in \mathcal{C}$ and $(\tau, \eta, \gamma) \in \mathbb{R} \times T^*_{\underline{x}}\partial\Omega \times]0, \infty[$. Introducing the fast variable z and scaling the frequency variables properly (see section 2 below), the Fourier-Laplace transform of (1.10) reads

$$\mathcal{L}v := -\widetilde{B}_n \frac{d^2 v}{dz^2} + A\frac{dv}{dz} + Mv \tag{1.14}$$

This is an ordinary differential system in z, depending on the parameters (b, u, ζ) with $\zeta := (\tau, \eta, \gamma)$.

Let $\mathbb{E}_-(b, u, \zeta)$ denote the set of initial data $(v(0), \frac{dv}{dz}(0)) \in \mathbb{C}^N \times \mathbb{C}^N$ such that the corresponding solution of $\mathcal{L}v = 0$ is bounded as z tends to infinity. Under Assumptions 1.1 and 1.2, for all $(b, u) \in \mathcal{C}$ and $\zeta = (\tau, \eta, \gamma) \neq 0$ with $\gamma \ge 0$, $\mathbb{E}_-(b, u, \zeta)$ has dimension N and depends smoothly on the parameters (b, u, ζ) (see Corollary 2.7 below). The *weak stability* condition states that the problem $\mathcal{L}v = 0$,

$v(0) = 0$ has no nontrivial bounded solutions, equivalently that $\mathbb{E}_-$ is transverse to $\ker \Gamma := \{0\} \times \mathbb{C}^N$, where Γ is the mapping $(\dot{u}, \dot{v}) \mapsto \dot{u}$ from $\mathbb{C}^N \times \mathbb{C}^N$ to $\mathbb{C}^N$. This is clearly necessary for linear stability, its violation implying the existence of local time-exponentially growing modes. The *strong or uniform* stability condition requires in addition some uniform behavior as ζ tends to zero and also as ζ tends to infinity. In particular, the uniform behavior near the origin is needed to recover the uniform stability of the hyperbolic boundary value problem. The uniform behavior at infinity is equivalent to the well-posedness of the Dirichlet boundary value problem for the parabolic part of the equation.

Because $\mathbb{E}_-$ and $\ker \Gamma$ both have dimension N in a space of dimension $2N$, there is a determinant

$$D(b, u, \zeta) = \det \big(\mathbb{E}_-(b, u, \zeta), \ker \Gamma \big) \tag{1.15}$$

obtained by taking orthonormal bases in each space, and the result is independent of the choice of the bases. This is the *Evans' function* (see [**Z**], [**S**]). D vanishes if and only if $\mathbb{E}_- \cap \ker \Gamma$ is not reduced to $\{0\}$.

To deal properly with the high frequencies, some appropriate scaling is required. With

$$\Lambda(\zeta) = \Big(1 + \tau^2 + \gamma^2 + |\eta|^4 \Big)^{\frac{1}{4}} \tag{1.16}$$

introduce the space $\widetilde{\mathbb{E}}_-(b, u, \zeta) = J_\Lambda \mathbb{E}_-(b, u, \zeta)$ where J_Λ is the mapping $(\dot{u}, \dot{v}) \mapsto (\dot{u}, \Lambda^{-1} \dot{v})$ in $\mathbb{C}^N \times \mathbb{C}^N$ and the "scaled" Evans' function

$$\widetilde{D}(b, u, \zeta) = \det \big(\widetilde{\mathbb{E}}_-(b, u, \zeta), \ker \Gamma \big) . \tag{1.17}$$

Note that $\ker \Gamma$ is invariant by J_Λ so that D vanishes if and only if $\widetilde{D}$ vanishes. Moreover, for bounded values of ζ, there is C such that $\frac{1}{C}|D| \le |\widetilde{D}| \le C|D|$, since, in the computation of the Evans's functions, the introduction of J_Λ only amounts to a change of scalar product in $\mathbb{C}^{2N}$.

The *weak* stability condition requires that $D \neq 0$ for $(b, u) \in \mathcal{C}$ and $\zeta \neq 0$ with $\gamma \ge 0$. The *strong* or *uniform* reads

ASSUMPTION 1.4 (Uniform stability condition). *There is a constant $c > 0$ such that for all for all $(b, u) \in \mathcal{C}$ and $\zeta = (\tau, \gamma, \eta) \neq 0$ with $\gamma \ge 0$*

$$|\widetilde{D}(b, u, \zeta)| \ge c \tag{1.18}$$

REMARKS 1.5. **a)** The weak and uniform stability conditions are conditions on the "frozen coefficient" planar boundary value problems associated with the inner layer solution. They are natural analogs of those defined in [**Z**] for the planar shock case. In the one-dimensional boundary layer case, Assumption 1.4 reduces to the condition imposed by Grenier and Rousset [**Gr-Ro**].

b) The uniform stability condition is equivalent to saying that

$$|v| \le C \Lambda |u| \quad \text{for} \quad U = {}^t(u, v) \in \mathbb{E}_-(b, u, \tau, \eta, \gamma) \tag{1.19}$$

uniformly with respect to (b, u) and (τ, η, γ) bounded, with $(\tau, \eta, \gamma) \neq 0$ and $\gamma \ge 0$.

c) Under Assumptions 1.1 and (1.2) the spaces $\mathbb{E}_-(b, u, \tau, \gamma, \eta)$ have limits $\mathbb{E}_-^0(b, u, \check{\tau}, \check{\eta}, \check{\gamma})$ when $(\tau, \gamma, \eta) = \rho(\check{\tau}, \check{\eta}, \check{\gamma})$ and ρ tends to zero, with $(\check{\tau}, \check{\eta}, \check{\gamma}) \neq 0$, $\check{\gamma} > 0$. In addition, the spaces $\mathbb{E}_-^0$ are closely related to the similar spaces associated to the limit hyperbolic problem, and extend continuously to $\check\gamma = 0$. The uniform stability condition implies that for all $(b, u) \in \mathcal{C}$ and $(\tau, \gamma, \eta) \neq 0$ with $\gamma \geq 0$:

$$\mathbb{E}_-(b, u, \zeta) \bigcap (\{0\} \times \mathbb{C}^N) = \{0\}, \tag{1.20}$$

and for all $(\hat\tau, \hat\gamma, \hat\eta) \neq 0$ with $\hat\gamma \geq 0$:

$$\mathbb{E}^0_-(b, u, \check\tau, \check\eta, \check\gamma) \bigcap (\{0\} \times \mathbb{C}^N) = \{0\}. \tag{1.21}$$

This will be shown in Appendix A.

d) The stability condition also involves a uniform behavior as (τ, η, γ) tends to infinity. Indeed, one can show that the spaces $\widetilde{\mathbb{E}}_-(y, b, u, \tau, \eta, \gamma)$ have limits as $\mathbb{E}^\infty_-(y, b, u, \tilde\tau, \tilde\gamma, \tilde\eta)$ when $(\tau, \gamma, \eta) = (\lambda^2\tilde\tau, \lambda^2\tilde\gamma, \lambda\tilde\eta)$ and λ tends to infinity, with $(\tilde\tau, \tilde\gamma, \tilde\eta) \neq 0$, $\tilde\gamma \geq 0$. The uniform stability condition (1.18) for large values of ζ is equivalent to the transversality condition $\mathbb{E}^\infty_- \cap \ker \Gamma = \{0\}$. It turns out that this condition is equivalent to the well-posedness of the *parabolic* Dirichlet boundary value problem, as can be seen by a standard rescaling/asymptotic ODE argument (see [**Z**], Lemma 4.28 and also the proof of Lemma 2.14 below). In particular, it is satisfied when the parabolic operator is symmetric, i.e. when there is a smooth definite positive $S(b, u)$ such that $\operatorname{Re}(\sum \xi_j \xi_k S B_{j,k})$ is positive definite for $\xi \neq 0$ (see Remark 2.15 below).

The following useful relation was established by Rousset [**Rou**] via Evans function calculations. A proof of the second part of the assertion (i.e., satisfaction of the uniform Kreiss-Lopatinski condition) is given in Appendix A; see also Remark c) above.

PROPOSITION 1.6. [**Rou**] *Under Assumptions* 1.1, 1.2, *the uniform stability condition Assumption* 1.4 *implies both that the limiting boundary condition* (1.8) *is transversal and that the resulting limiting hyperbolic boundary value problem* (1.1) (1.8) *satisfies the uniform Kreiss-Lopatinski condition. (Indeed, these two statements are together equivalent to Assumption* 1.4(*ii*).*)*

Therefore, under Assumptions 1.1, 1.2 and 1.4, one can solve the mixed problem (1.1) (1.8) for initial conditions which satisfy sufficiently many compatibility conditions (see [**Maj**], [**Ra-Ma**], [**Mok**], [**Mé2**]).

REMARK 1.7. For outer initial data sufficiently close to some particular value u_0 for which there exists a boundary layer satisfying the uniform stability condition 1.4 (or, more generally, a compact set $\mathcal{U}_0$ of such values), Remark 1.5 (b) above, together with the above proposition, implies that Assumption 1.4 is automatically satisfied for such time as the outer solution remains near u_0 (resp. $\mathcal{U}_0$). This gives a simple situation in which these assumptions are verifiable a priori for solutions with large boundary layer. Note that this does not preclude the possibility of multiple (but locally unique), stable boundary layers, with associated distinct valid boundary layer expansions. For analogs in the shock layer setting, see [**AMPZ**].

Consider a solution u_0 in $H^{s_0}([-T_0, T_0] \times \Omega)$ of the hyperbolic boundary value problem (1.1) (1.8), with b a given smooth enough function in $H^{s_0}([-T_0, T_0] \times \Omega)$. The index s_0 is large enough, and how large will be made precise later. In any case, $s_0 > 1 + \frac{d+1}{2}$ so that functions in H^{s_0} are Lipschitz continuous. By definition of the boundary condition, there is a profile

$$w_0(t, y, z) = w_{b(t,y), u_0(t,y)}(z). \tag{1.22}$$

The profiles $w_{b,a}$ are defined for $b \in \mathcal{B}_\partial$ and $a \in \mathcal{C}_b$. It is convenient to extend the definition to all b and $a \in \mathcal{U}$.

LEMMA 1.8. *There is a C^∞ function W on $\mathcal{O} \times [0, \infty[$ such that*
i) $W(b, a, z) = 0$ when $b \in \mathcal{B}_\partial$ and $z = 0$,
ii) for all compact set $\mathcal{K} \subset \times\mathcal{U}$, there are $\delta > 0$ and C such that

$$\forall b \in \mathcal{B}\,,\ \forall a \in \mathcal{K}\,, \forall z \geq 0\,: \quad |W(b, a, z) - a| \leq Ce^{-\delta z}$$

iii) when $b \in \mathcal{B}_\partial$ and $a \in \mathcal{C}_b$ then $z \mapsto W(b, a, z)$ is a solution of (1.6)

PROOF. One can parametrize a neighborhood of $\partial\Omega$ using normal coordinates $x = y + x_n\nu(y)$. Near $(\underline{b}, \underline{a}) \in \mathcal{C}$, one can use coordinates $a = (a', a'')$ such that $\mathcal{C}$ is given by the equations $a'' = h(b, a')$. Then one can extend locally the function w as

$$W(b, a, z) = w_{b^\flat, a', h(b^\flat, a')}(z) + (a'' - h(b^\flat, a')) \tanh z\,.$$

where $b^\flat = (t, y, b_0)$ if $b = (t, x, b_0)$ and $x = y + x_n\nu(y)$. When x is outside a neighborhood of $\partial\Omega$, one can take $W = a$. Gluing the pieces by a partition of unity yields the result. □

Taking $\varphi \in C^\infty(\overline{\Omega})$ such that $\varphi = 0$ and $d\varphi = \nu$ on $\partial\Omega$, introduce

$$u_0^\varepsilon(t, x) = W(b(t, x), u_0(t, x), \varphi(x)/\varepsilon)\,. \tag{1.23}$$

By construction, it satisfies

$$\begin{cases} u^\varepsilon_{0|\partial\Omega} = 0\,. \\ u_0^\varepsilon - u_0 \doteq O(e^{-\delta\varphi/\varepsilon}) \end{cases} \tag{1.24}$$

Thus u_0^ε is a perturbation of u_0 in the interior and the general idea is that u_0^ε is close to a solution of (1.4). In this direction, the main step is to prove that the linearized equations from (1.4) around u_0^ε are stable. More precisely, consider a family of perturbations $v^\varepsilon \in W^{1,\infty}([-T_0, T] \times \Omega)$ and the linearized equation from (1.4) around $u_a^\varepsilon := u_0^\varepsilon + \varepsilon v^\varepsilon$:

$$\mathcal{P}_{u_0^\varepsilon + \varepsilon v}(t, x, \partial_t, \partial_x)u = f\,, \quad u_{|\partial\Omega} = 0 \tag{1.25}$$

$\mathcal{P}_{\tilde{u}_0^\varepsilon + \varepsilon v}$ is a differential operator, first order in t and second order in x, whose coefficients depend on b, $\tilde{u}_0^\varepsilon$ and v and their derivatives. Its explicit form is computed in section 4 below.

The main new result of this paper is that, under Assumptions 1.1 and 1.4, the equations (1.25) are well posed in L^2. With $T_0 > 0$ given, we assume that

$$\begin{cases} u_0 \in W^{2,\infty}([-T_0, T_0] \times \Omega)\,, \quad b \in W^{2,\infty}([-T_0, T_0] \times \Omega) \\ \sup\limits_{\varepsilon \in]0,1]} \Big(\|v^\varepsilon\|_{L^\infty} + \|\varepsilon \nabla_{t,x} v^\varepsilon\|_{L^\infty} + \|\varepsilon^2 \nabla^2_{t,x} v^\varepsilon\|_{L^\infty} \Big) < \infty\,. \end{cases} \tag{1.26}$$

THEOREM 1.9 (L^2 stability). *There are $C > 0$ and ε_0 such that for all $\varepsilon \in]0, \varepsilon_0]$ and $f \in L^2([-T_0, T_0] \times \Omega)$ vanishing for $t < 0$, the equation* (1.25) *has a unique solution which vanishes for $t < 0$. Moreover*

$$\|u\|_{L^2} + \sqrt{\varepsilon}\|\partial_x u\|_{L^2} + \varepsilon^{3/2}\|\partial_x^2 u\|_{L^2} \leq C\|f\|_{L^2}\,. \tag{1.27}$$

This theorem is proved in section 4 below together with slight improvements which are needed in the proof of the estimates for the derivatives. Let us just mention where the difficulty lies. The coefficients of $\mathcal{P}_{\tilde{u}_0^\varepsilon}$ depend on $\varphi(x/\varepsilon)$ and thus are not (uniformly) Lipschitzean. Moreover, the coefficient of u in $\mathcal{P}_{u_a^\varepsilon}$ has a factor $\frac{1}{\varepsilon}$ in front of it. Thus the usual energy method using integration by parts yields singular and apparently uncontrolled terms. This is exactly where the smallness assumption in [**Gr-Gu**] comes in. Using it together with a tricky argument, the authors were able to absorb the singular terms. Our main objective in this paper, is to replace the smallness argument by a detailed analysis of $\mathcal{P}_{u_a^\varepsilon}$ and to use the Assumption 1.4 to construct symmetrizers.

The next step is to prove estimates for the derivatives of the solution u. The classical approach is to differentiate (1.25) with respect to vector fields which are tangent to $\mathbb{R} \times \partial\Omega$, in order have natural boundary conditions for the derivatives. For non characteristic problems, the normal derivatives are deduced from the tangential ones using the equations. Here, we can adapt the first argument, but the second certainly fails since the coefficients of $\mathcal{P}_{u_a^\varepsilon}$ are singular in the normal direction and the solution cannot be (uniformly) smooth in this direction. This leads us to introduce spaces with *conormal* Sobolev smoothness. Such spaces have already been widely used in the study of boundary value problems, see e.g. [**Ra2**], [**Gu2**]. Let $\{Z_k\}_{0\leq k\leq \underline{k}}$ denote a finite set of generators of vector fields tangent to $\mathbb{R} \times \partial\Omega$, with for instance $Z_0 = \partial_t$. For $U \subset \mathbb{R} \times \Omega$ and $m \in \mathbb{N}$, define the space

$$(1.28)\qquad \begin{aligned} \mathcal{H}^m(U) := \big\{ u \in L^2(U) \, : Z_{k_1} \dots Z_{k_p} u \in L^2(U)\,, \\ \forall p \leq m \,, \forall (k_1, \dots, k_p) \in \{0, \dots \underline{k}\}^p \,\big\} \end{aligned}$$

This space is equipped with the obvious norm, denoted by $\| \cdot \|_{\mathcal{H}^m(U)}$.

In order to solve nonlinear problems, we need work in Banach algebras which means here that we have to supplement the $\mathcal{H}^m$ estimates with L^∞ estimates. Introduce the following norms

$$(1.29)\qquad \|u\|_{\mathcal{W}^\mu(U)} = \|u\|_{L^\infty} + \sum_{p=1}^{\mu} \sum_{1\leq k_1,\dots,k_p\leq \underline{k}} \|Z_{k_1} \dots Z_{k_p} u\|_{L^\infty} \,.$$

Reinforcing (1.26), we now assume that on $\Omega_{T_0} := [-T_0, T_0] \times \Omega$,

$$(1.30)\qquad \begin{cases} u_0 \in W^{m+2,\infty}(\Omega_{T_0})\,, \qquad b \in W^{m+2,\infty}(\Omega_{T_0}) \\ \sup\limits_{\varepsilon\in]0,1]} \|v^\varepsilon\|_{\mathcal{W}^m} + \varepsilon \|\nabla_{t,x} v^\varepsilon\|_{\mathcal{W}^m} + \varepsilon^2 \|\nabla_x^2 v^\varepsilon\|_{\mathcal{W}^m} < \infty \,. \end{cases}$$

THEOREM 1.10. *There are $C > 0$ and ε_0 such that all $\varepsilon \in]0, \varepsilon_0]$ and all $f \in \mathcal{H}^m([-T_0, T_0] \times \Omega)$ vanishing for $t < 0$, the solution of equation* (1.25) *satisfies*

$$(1.31)\qquad \|u\|_{\mathcal{H}^m} + \sqrt{\varepsilon}\|\partial_x u\|_{\mathcal{H}^m} + \varepsilon^{3/2}\|\partial_x^2 u\|_{\mathcal{H}^m} \leq C\|f\|_{\mathcal{H}^m}$$

If in addition $m \geq 2 + \frac{d+1}{2}$ and $f \in L^\infty([-T_0, T_0] \times \Omega)$, then the solution u also satisfies

$$(1.32)\qquad \|u\|_{\mathcal{W}^2} + \varepsilon\|\partial_x u\|_{\mathcal{W}^1} + \varepsilon^2\|\partial_x^2 u\|_{L^\infty} \leq C\big(\|f\|_{\mathcal{H}^m} + \varepsilon\|f\|_{L^\infty}\big)\,.$$

These results can be used to solve the nonlinear equations (1.4). In order to avoid technical discussions on compatibility conditions for the Cauchy data and the boundary conditions, we consider here the simple case where the Cauchy data for (1.1) and (1.2) are zero, but with a non trivial forcing term, see [**Gr-Gu**]. More precisely, we consider $F(b,u)$ such that $F(0,0) = 0$. With indices m and s_0 such that

$$m > \frac{d+1}{2}, \qquad s_0 > m + 3\frac{d+1}{2}, \tag{1.33}$$

consider $b \in H^{s_0}([-T_0,T_0]\times\Omega)$ such that $b = 0$ for $t < 0$. Assuming that the state $u = 0$ belongs the domain of hyperbolicity $\mathcal{O}$ in Assumption 1.1 and shrinking T_0 if necessary, the mixed Cauchy problem (1.1) (1.8) has a unique solution $u_0 \in H^{s_0}([-T_0,T_0]\times\Omega)$ which vanishes for $t < 0$. In this case, u_0^ε given by (1.23) vanishes for $t < 0$ and is an exact solution of (1.4) there. We show that this solution can be continued to $[0,T_0]\times\Omega$ and that u_0^ε is a good approximation.

THEOREM 1.11. *There is $\varepsilon_0 > 0$ such that for all $\varepsilon \in]0,\varepsilon_0]$ the problem* (1.4)(1.5) *has a unique solution u^ε which vanishes for $t < 0$. Moreover,*

$$\|u^\varepsilon - u_0^\varepsilon\|_{\mathcal{H}^m} + \|u - u_0^\varepsilon\|_{L^\infty} = O(\varepsilon). \tag{1.34}$$

This theorem is proved in section 6. Indeed, we first construct a first corrector u_1^ε such that $u_1^\varepsilon = 0$ for $t < 0$, $u_1^\varepsilon = 0$ on $[-T_0,T_0]\times\partial\Omega$ and $u_a^\varepsilon = u_0^\varepsilon + \varepsilon u_1^\varepsilon$ satisfies equation (1.4) up to an error $e = O(\varepsilon)$. Indeed, when one substitutes u_0^ε in (1.4), the $O(\varepsilon^{-1})$ term is killed by the choice (1.23) and because W satisfies (1.6) when the boundary condition is satisfied. However, it remains an $O(e^{-\delta\varphi/\varepsilon})$ term. The corrector u_1^ε, given by a formula analogous to (1.23), can be chosen to cancel this term (see the general discussion of BKW solutions in [**Gr-Gu**]). Then the solution u^ε is constructed as $u_a^\varepsilon + \varepsilon v^\varepsilon$, where v^ε solves

$$\mathcal{P}_{u_a^\varepsilon} v^\varepsilon + \varepsilon\mathcal{Q}(v^\varepsilon) = f := \varepsilon^{-1} e \tag{1.35}$$

and $\mathcal{Q}$ is at least quadratic in v. Denoting by $\|\cdot\|_{\mathcal{X}_\varepsilon^m}$ [resp. $\|\cdot\|_{\mathcal{Y}_\varepsilon^m}$] the norm given by adding the left [resp right] hand sides of (1.31) and (1.32) one proves that

$$\begin{aligned} &\|\varepsilon\mathcal{Q}(v^\varepsilon)\|_{\mathcal{Y}_\varepsilon^m} \le \varepsilon^{1/4}C(M), \\ &\|\varepsilon(\mathcal{Q}(v_1^\varepsilon) - \mathcal{Q}(v_2^\varepsilon))\|_{\mathcal{Y}_\varepsilon^m} \le \varepsilon^{1/4}C(M)\,\|v_1 - v_2\|_{\mathcal{X}_\varepsilon^m}, \end{aligned} \tag{1.36}$$

provided that

$$\begin{aligned} &\varepsilon\|v_1\|_{L^\infty} \le 1, \qquad &\varepsilon\|v_1\|_{L^\infty} \le 1 \\ &\varepsilon\|v_1\|_{\mathcal{X}_\varepsilon^m} \le M, \qquad &\varepsilon\|v_1\|_{\mathcal{X}_\varepsilon^m} \le M. \end{aligned} \tag{1.37}$$

Together with Theorem 1.10, this shows that the equation (1.35) can be solved in $\mathcal{X}^m$, provided that ε is small enough.

The main result in [**Gr-Gu**] is analogous to Theorem 1.11, but does not give the existence up to T_0. They proved the linear and nonlinear stability as long as u_0^ε remains smaller than some small constant in a suitable norm. Here, we get the stability under the more geometric condition that $(y, u_0(t,y))$ remains in a domain where the uniform stability condition Assumption 1.4 holds, knowing that when this condition fails, strong instabilities can occur. Note also that our estimates in Theorem 1.10 are stronger than the corresponding estimates in [**Gr-Gu**], since they proved estimates for derivatives εZ and not for Z. The price they had to pay was that they needed a very accurate approximate solution u_a^ε, so that the solution is constructed as $u_a^\varepsilon + \varepsilon^M v$ with M large, so that control of ε derivatives for v gives control of L^∞ norms for $\varepsilon^M v$. Moreover, the accurate approximate solutions were constructed by BKW expansions, which require a lot of smoothness on u_0. Indeed, they assumed $u_0 \in C^\infty$.

However, the results in Theorem 1.10 and 1.11 are not quite satisfactory. Because (1.4) is parabolic, one should expect the solutions to be smoother than the solutions of (1.1). Here we get a result going the wrong way. We start from a very smooth solution u_0 of (1.1) and we end up with less smooth solutions of (1.4). This is clearly related to the method of proof, and a direct proof of existence with uniform estimates for (1.4), without using the solution u_0 of (1.1) would be very interesting. This will be developed in a further work. In any case, the stability analysis in Theorem 1.9 is the key point and is indeed the main concern of this paper.

2. Linear stability: the model case

This section is an introduction to and preparation for the general analysis developed in section 4. We consider the model problem where the domain is a half plane and the hyperbolic solution u_0 as well as the source term are constant. More precisely, we fix $\underline{y} \in \partial\Omega$ and a local chart χ from a neighborhood $\mathcal{V}$ of $\underline{y}$ to a neighborhood $\widetilde{\mathcal{V}}$ of 0 in $\mathbb{R}^d$ such that $\Omega \cap \mathcal{V}$ is transformed into $\widetilde{\mathcal{V}} \cap \{x > 0\}$ where $(y, x) \in \mathbb{R}^{d-1} \times \mathbb{R}$ are coordinates in $\mathbb{R}^d$, and the defining function φ in (1.23) is $\varphi = x$. The form of the equation (1.4) is preserved by the change of coordinates, as well as the Assumptions 1.1 and 1.4. For simplicity, we keep the same notations. The linearized equations around u_a^ε read

$$(2.1)\qquad \begin{cases} \partial_t u + \displaystyle\sum_{j=1}^{d} A_j^\varepsilon \partial_j u - \varepsilon \sum_{j,k=1}^{d} B_{j,k}^\varepsilon \partial_{j,k}^2 u + \frac{1}{\varepsilon} E^\varepsilon u = f\,, \qquad x > 0\,, \\ u_{x=0} = 0\,. \end{cases}$$

with

$$\begin{cases} A_j^\varepsilon v = \widetilde{A}_j v - \varepsilon \displaystyle\sum_k (\widetilde{B}'_{j,k} \cdot v) \partial_k u_a^\varepsilon - \varepsilon \sum_k (\widetilde{B}'_{k,j} \cdot \partial_k u_a^\varepsilon) v\,, \\ B_{j,k}^\varepsilon = \widetilde{B}_{j,k}\,, \\ E^\varepsilon v = \varepsilon \displaystyle\sum_j (\widetilde{A}'_j \cdot v) \partial_j u_a^\varepsilon - \varepsilon^2 \sum_{j,k} (\widetilde{B}'_{j,k} \cdot v) \partial_{j,k}^2 u_a^\varepsilon + \widetilde{B}''_{j,k}(v, \partial_j u_a^\varepsilon) \partial_k u_a^\varepsilon\,, \end{cases}$$

where $\widetilde{A}$ stands for the function A evaluated at $(b(t, y, x), u_a^\varepsilon(t, y, x))$ and A' is the derivative of A with respect to the variable u. The $d - 1$ first variables are $(y_1, \ldots, y_{d-1})$ and the d-th variable is x.

In particular, we will consider (2.1) when $u_a^\varepsilon = u_0^\varepsilon + \varepsilon v$ and u_0^ε is given by (1.23), which reads in the new coordinates,

$$(2.2)\qquad u_0^\varepsilon(t, y, x) = W(b(t, y, x), u_0(t, y, x), \frac{x}{\varepsilon})$$

As mentioned in the introduction, the starting point is to analyze the linearized equation when u_0 is a constant and the coefficients (t, y, x) are frozen. This leads us to consider functions

$$(2.3)\qquad u_a^\varepsilon(t, y, x) = W(b, a, \frac{x}{\varepsilon}) + c := w(p, \frac{x}{\varepsilon})$$

where

$$(2.4)\qquad p = (b, a, c)$$

are parameters which vary in a neighborhoods of $\underline{b} = b(\underline{t}, 0, 0)$ in $\mathbb{R} \times \mathbb{R}^d \times \mathbb{R}^{M_0} \times [0, \infty[$, $\underline{a} = u_0(\underline{t}, 0, 0)$ in $\mathcal{O}$, and $\underline{c} = 0$ in $\mathbb{R}^N$ respectively. We denote by $\underline{p} = (\underline{b}, \underline{a}, \underline{c})$.

When u_a^ε is given by (2.2), the linearized equation (2.1) simplifies to

$$\tag{2.5} \begin{cases} \partial_t u + \displaystyle\sum_{j=1}^d A_j^\sharp \partial_j u - \varepsilon \sum_{j,k=1}^d B_{j,k}^\sharp \partial_{j,k}^2 u + \frac{1}{\varepsilon} E^\sharp u = f\,, & x > 0\,, \\ u_{x=0} = 0\,. \end{cases}$$

with

$$\tag{2.6} \begin{cases} A_j^\sharp v = \widetilde{A}_j v - (\widetilde{B}'_{j,d} \cdot v)\partial_z w_p - (\widetilde{B}'_{d,j} \cdot \partial_z w_p) v\,, \\ B_{j,k}^\sharp = \widetilde{B}_{j,k}\,, \\ E^\sharp v = (\widetilde{A}'_d \cdot v)\partial_z w_p - (\widetilde{B}'_{d,d} \cdot v)\partial_z^2 w_p - \widetilde{B}''_{d,d}(v, \partial_z w_p)\partial_z w_p\,, \end{cases}$$

and the functions are now evaluated at $(b, w(p, x/\varepsilon))$. We remark that all the coefficients $A_j^\sharp$, $B_{j,k}^\sharp$ and $E^\sharp$ are C^∞ functions of p and $z = x/\varepsilon$. Moreover, they converge with an exponential rate when z tends to $+\infty$. The limits are denoted by $A_j^\infty(p)$, $B_{j,k}^\infty(p)$ and $E^\infty(p)$. They are given by

$$\tag{2.7} A_j^\infty(p) = A_j(b, a + c)\,, \quad B_{j,k}^\infty(p) = B_{j,k}(b, a + c)\,, \quad E^\infty(p) = 0\,.$$

Moreover, there are C and $\delta > 0$, such that for all p in a neighborhood of $\underline{p}$ and all indices j and k:

$$\tag{2.8} |A_j^\sharp(p, z) - A^\infty(p)| + |B_{j,k}^\sharp(p, z) - A^\infty(p)| + |E^\sharp(p, z)| \le Ce^{-\delta z}\,.$$

In this section we prove uniform *a priori* estimates for (2.5). We shall concentrate more on the method than on the estimates themselves. The symmetrizers we construct now will serve as *symbols* for the general construction of symmetrizers performed in section 4.

We denote by $\|\cdot\|$ and $|\cdot|$ the L^2 norms respectively on the half space $\{(t, y, x) \in \mathbb{R}^{1+d} : x > 0\}$ and on the boundary $\{(t, y) \in \mathbb{R}^d\}$.

THEOREM 2.1. *There is C such that for all $\varepsilon \in]0, 1]$, all $\gamma > 0$ and all test functions u, f satisfying* (2.1), *one has*

$$\tag{2.9} \begin{aligned} &\gamma\|e^{-\gamma t} u\| + \sqrt{\varepsilon\gamma}\|e^{-\gamma t}\partial_y u\| + \sqrt{\varepsilon\gamma}\|e^{-\gamma t}\partial_x u\| \\ &\qquad + \varepsilon\|e^{-\gamma t}\partial_y^2 u\| + \varepsilon\|e^{-\gamma t}\partial_y\partial_x u\| \le C\|e^{-\gamma t} f\| \end{aligned}$$

The trace $\partial_x u_{|x=0}$ can also be estimated as stated in section 4, but this is unessential here. Introducing $\tilde{u} = e^{-\gamma t} u$, (2.5) is equivalent to

$$\tag{2.10} \begin{cases} (\partial_t + \gamma)\tilde{u} + \displaystyle\sum_{j=1}^d A_j^\sharp \partial_j \tilde{u} - \varepsilon \sum_{j,k=1}^d B_{j,k}^\sharp \partial_{j,k}^2 \tilde{u} + \frac{1}{\varepsilon} E^\sharp \tilde{u} = \tilde{f}\,, \\ \tilde{u}_{x=0} = 0\,. \end{cases}$$

with $\tilde{f} = e^{-\gamma t} f$. Thus (2.9) is equivalent to

$$\tag{2.11} \gamma\|\tilde{u}\| + \sqrt{\varepsilon\gamma}\|\partial_{y,x}\tilde{u}\| + \varepsilon\|\partial_{y,x}\partial_y \tilde{u}\| \le C\|\tilde{f}\|$$

for the solutions of (2.10) We write (2.10) in the condensed form

$$\tag{2.12} \begin{cases} -\varepsilon\partial_x^2 \tilde{u} + A^\sharp \partial_x \tilde{u} + \dfrac{1}{\varepsilon} M^\sharp \tilde{u} = (B_{d,d}^\sharp)^{-1} \tilde{f}\,, \\ u_{x=0} = 0\,. \end{cases}$$

where

$$\begin{cases} A^\sharp = (B^\sharp_{d,d})^{-1}\Big(A^\sharp_1 - \sum_{j=1}^{d-1}(B^\sharp_{j,d} + B^\sharp_{d,j})\varepsilon\partial_j\Big) \\ M^\sharp = (B^\sharp_{d,d})^{-1}\Big(\varepsilon(\partial_t+\gamma) + \sum_{j=1}^{d-1} A^\sharp_j\varepsilon\partial_j - \sum_{j,k=1}^{d-1} B^\sharp_{j,k}\varepsilon^2\partial^2_{j,k} + E^\sharp\Big). \end{cases}$$

We have used that, by Assumption (H1), $B_{d,d}$ and thus $B^\sharp_{d,d}$ are invertible. Remark that $A^\sharp$ and $M^\sharp$ are differential operators in $\varepsilon(\partial_t+\gamma)$ and $\varepsilon\partial_y$. They are respectively of order 1 and of order two for the natural parabolic weights, where $\varepsilon\partial_y$ has weight 1 and $\varepsilon\partial_t$ and $\varepsilon\gamma$ have weight 2.

Write (2.12) as a first order system for $U = \begin{pmatrix}\tilde u\\ \varepsilon\partial_x\tilde u\end{pmatrix}$:

$$\begin{cases} \partial_x U - \dfrac{1}{\varepsilon}G^\sharp U = F\,, \\ \Gamma U_{x=0} = 0\,. \end{cases} \tag{2.13}$$

with

$$\Gamma\begin{pmatrix}u\\v\end{pmatrix} = u \quad G^\sharp = \begin{pmatrix}0 & \mathrm{Id}\\ M^\sharp & A^\sharp\end{pmatrix}, \qquad F = \begin{pmatrix}0\\ -(B^\sharp_{d,d})^{-1}\tilde f\end{pmatrix}.$$

Theorem 2.1 follows from the following estimates for the solutions of (2.13): there is C such that for all $\varepsilon\in]0,\varepsilon_0]$, all γ_0 such that for all $\gamma\geq\gamma_0$ and all test functions $U=\begin{pmatrix}u\\v\end{pmatrix}$, F satisfying (2.13), one has

$$\gamma\|u\| + \sqrt{\varepsilon\gamma}\|\partial_y u\| + \varepsilon\|\partial_y^2 u\| + \frac{\sqrt{\gamma}}{\sqrt{\varepsilon}}\|v\|^2 + \|\partial_y v\| \leq C\|F\|^2 \tag{2.14}$$

2.1. Symmetrizers. Recall now the essence of the "method of symmetrizers" as it applies to general boundary value problems

$$\partial_x u = G(x)u + f\,, \quad \Gamma u(0) = 0\,. \tag{2.15}$$

Here, u and f are functions on $[0,\infty[$ values in some Hilbert space $\mathcal{H}$, and $G(x)$ is a C^1 family of (possibly unbounded) operators defined on $\mathcal{D}$, dense subspace of $\mathcal{H}$.

A symmetrizer is a family of C^1 functions $x\mapsto S(x)$ with values in the space of operators in $\mathcal{H}$ such that there are C_0, $\lambda>0$, $\delta>0$ and C_1 such that

$$\forall x\,,\quad S(x)=S(x)^* \quad\text{and}\quad |S(x)|\leq C_0\,, \tag{2.16}$$

$$\forall x\,,\quad 2\mathrm{Re}\,S(x)G(x) + \partial_x S(x) \geq 2\lambda\mathrm{Id}\,, \tag{2.17}$$

$$S(0)\geq \delta Id - C_1\Gamma^*\Gamma\,. \tag{2.18}$$

In (2.16), the norm of $S(x)$ is the norm in the space of bounded operators in $\mathcal{H}$. Similarly $S^*(x)$ is the adjoint operator of $S(x)$. The notation $\mathrm{Re}\,T = \frac{1}{2}(T+T^*)$ is used in (2.17) for the real part of an operator T. When T is unbounded, the meaning of $\mathrm{Re}\,T\geq\lambda$, is that all $u\in\mathcal{D}$ belongs to the domain of T and satisfies

$$\mathrm{Re}\,\big(Tu,u\big)\geq\lambda|u|^2\,. \tag{2.19}$$

The property (2.17) has to be understood in this sense.

LEMMA 2.2. *If there is a symmetrizer S, then for all $u \in C_0^1([0,\infty[;\mathcal{H}) \cap C^0([0,\infty[;\mathcal{D})$, one has*

$$\lambda\|u\|^2 + \delta|u(0)|^2 \leq \frac{C_0^2}{\lambda}\|f\|^2 + C_1|\Gamma u(0)|^2 , \tag{2.20}$$

where $f := \partial_x u - Gu$.

Here, $|\cdot|$ is the norm in $\mathcal{H}$ and $\|\cdot\|$ the norm in $L^2([0,\infty[;\mathcal{H})$.

PROOF. Taking the scalar product of Su with the equation (2.15) and integrating over $[0,\infty[$, (2.16) implies

$$\begin{aligned} -(S(0)u(0),u(0)) &= \int \partial_x(Su,u)dx \\ &= \int ((2\mathrm{Re}\, SG + \partial_x S)u,u)dx + 2\mathrm{Re} \int (Sf,u)dx \,. \end{aligned} \tag{2.21}$$

By (2.17),

$$\int ((2\mathrm{Re}\, SG + \partial_x S)u,u)dx \geq 2\lambda\|u\|^2 \,.$$

By (2.18) and the boundary condition $\Gamma u(0) = 0$,

$$(S(0)u(0),u(0)) \geq \delta|u(0)|^2 - C_1|\Gamma u(0)|^2 \,.$$

By (2.16)

$$2\Big|\int (Sf,u)dx\Big| \leq 2C_0\|f\|\,\|u\| \leq \frac{C_0^2}{\lambda}\|f\|^2 + \lambda\|u\|^2 \,.$$

Thus the identity (2.21) implies the energy estimate (2.20). □

2.2. Laplace–Fourier transform. To obtain energy estimates for (2.13) we perform a Fourier transform in variables (t,y). The matrix $G^\sharp$ in (2.13) is a differential operator in $\varepsilon\partial_t, \varepsilon\partial_y$ with coefficients independent of (t,y). Denoting by $\hat{U}(\tau,\eta,x)$ the Fourier transform of $U(t,y,x)$, (2.13) is equivalent to

$$\partial_x \hat{U} = \frac{1}{\varepsilon}\mathcal{G}(\frac{x}{\varepsilon},p,\varepsilon\tau,\varepsilon\gamma,\varepsilon\eta)\hat{U} + F\,, \quad \Gamma\hat{U}_{|x=0} = \hat{u}_{|x=0} = 0\,. \tag{2.22}$$

where

$$\mathcal{G}(z,p,\tau,\gamma,\eta) = \begin{pmatrix} 0 & \mathrm{Id} \\ \mathcal{M} & \mathcal{A} \end{pmatrix},$$

$$\begin{cases} \mathcal{A}(z,p,\tau,\gamma,\eta) = (B_{d,d}^\sharp)^{-1}\Big(A_1^\sharp - \displaystyle\sum_{j=1}^{d-1} i\eta_j(B_{j,d}^\sharp + B_{d,j}^\sharp)\Big) \\ \mathcal{M}(z,p,\tau,\gamma,\eta) = (B_{d,d}^\sharp)^{-1}\Big((i\tau+\gamma) + \displaystyle\sum_{j=1}^{d-1} i\eta_j A_j^\sharp + \sum_{j,k=1}^{d-1}\eta_j\eta_k B_{j,k}^\sharp + E^\sharp\Big)\,. \end{cases}$$

and the functions $A_j^\sharp$ etc are evaluated at (p,z).

By Plancherel's theorem, the energy estimates (2.14) are equivalent to the following estimates for the solutions of (2.22):

$$(\gamma + \varepsilon|\eta|^2)\|\hat{u}\| + (\frac{\sqrt{\gamma}}{\sqrt{\varepsilon}} + |\eta|)\|\hat{v}\| \leq C\|\hat{F}\|. \tag{2.23}$$

It is convenient to eliminate the ε in (2.22) by setting

$$\widetilde{U}(z) = \hat{U}(\varepsilon z)\,, \quad \widetilde{F}(z) = \varepsilon\hat{F}(\varepsilon z)\,, \quad (\tilde{\tau},\tilde{\gamma},\tilde{\eta}) = (\varepsilon\tau,\varepsilon\gamma,\varepsilon\eta)\,. \tag{2.24}$$

Then, (2.22) is transformed into

$$\partial_z \widetilde{U} = \mathcal{G}(z, p, \tilde{\tau}, \tilde{\gamma}, \tilde{\eta})\widetilde{U} + \widetilde{F}, \quad \Gamma\widetilde{U}(0) = u(0) = 0. \tag{2.25}$$

With (2.24), the energy estimate (2.23) is equivalent to the energy estimate

$$(\tilde{\gamma} + |\tilde{\eta}|^2)\|\tilde{u}\|^2 + (\sqrt{\tilde{\gamma}} + |\tilde{\eta}|)\|\tilde{v}\| + \leq \frac{C}{\gamma}\|\widetilde{F}\| \tag{2.26}$$

for the solutions of (2.25).

We now proceed to the proof of (2.26) and to the analysis of (2.25). We denote by $\zeta = (\tilde{\tau}, \tilde{\gamma}, \tilde{\eta}) \in \mathbb{R} \times [0, \infty[\times \mathbb{R}^{d-1}$ the frequencies. There are three different regimes, depending on the size of $|\zeta|$:

1) $|\zeta|$ is small. (Remember that we will perform the substitution (2.24), so that in the original variables this means that $\varepsilon(|\tau| + |\gamma| + |\eta|)$ is small). In this regime, part of u is governed by an elliptic-parabolic equation and the other part is governed by the limiting hyperbolic problem.

2) $|\zeta| \approx 1$. This is the intermediate regime, where (2.25) has an "elliptic" behavior.

3) $|\zeta|$ is large. In this high frequency regime, the parabolic behavior prevails. There is a natural quasi-homogeneity, where η has weight 1 and (τ, γ) have weight 2. The length is given by $\langle\zeta\rangle = (\tilde{\tau}^2 + \tilde{\gamma}^2 + |\tilde{\eta}|^4)^{1/4}$.

The next result summarizes the estimates in the different regimes. It implies (2.26) and therefore Theorem 2.1. Introduce a weight function $h(\zeta)$ such that

$$h(\zeta) \approx \begin{cases} (\tilde{\gamma} + |\zeta|^2)^{1/2} & \text{when } |\zeta| \leq 1, \\ \langle\zeta\rangle & \text{when } |\zeta| \geq 1. \end{cases} \tag{2.27}$$

Note that both $(\tilde{\gamma} + |\zeta|^2)^{1/2}$ and $\langle\zeta\rangle$ are ≈ 1 when $|\zeta| \approx 1$. Introduce next

$$\ell(\zeta) = h(\zeta)(1 + \langle\zeta\rangle)^{-1/2}, \tag{2.28}$$

which is of order h when $|\zeta| \leq 1$ and of order $\langle\zeta\rangle^{1/2}$ when $|\zeta| \geq 1$.

THEOREM 2.3. *There are a neighborhood of $\underline{p}$ and a constant C such that for all p in this neighborhood, all $\zeta \in \mathbb{R}^{d+1}$ with $\tilde{\gamma} > 0$ and all $\widetilde{U}$ and $\widetilde{F}$ in $C_0^\infty([0, \infty[)$ satisfying* (2.25), *one has*

$$h^2\|\tilde{u}\| + h\|\tilde{v}\| + \ell|\tilde{v}(0)| \leq C\|F\|. \tag{2.29}$$

REMARK 2.4. The inequality (2.29) implies the desired estimates (2.26)$\Leftrightarrow$ (2.23)$\Leftrightarrow$ (2.14) $\Rightarrow$ (2.9) but is much more precise. It includes estimates of the traces for $v = \varepsilon\partial_x u$ and also estimates for $\sqrt{\varepsilon}\partial_t u$ since $h \geq \min(1, \sqrt{\gamma})\,\tau$.

2.3. Reduction to constant coefficients. We prove Theorem 2.3 by constructing symmetrizers in the three different regimes. Remember that the frequencies for (2.25) are smaller by a factor ε than the actual frequencies for (2.22) (see (2.24)), so that the regime $\zeta \to 0$ is crucial. The main idea is to conjugate system (2.25), for bounded frequencies ζ, to a constant coefficient system

$$\partial_z U_1 = \mathcal{G}^\infty U_1 + F_1, \quad \Gamma_1 U_1(0) = 0, \tag{2.30}$$

using the exponential convergence of the coefficients at $z = \infty$. Thus we are reduced to constant coefficient equations and we perform the classical analysis, looking at the growing and decaying modes of $\mathcal{G}^\infty$. Here we extend Kreiss' analysis to parabolic-hyperbolic systems (2.30). The case where $|\zeta| \geq \rho_0 > 0$ falls in the so called elliptic region. The hyperbolic behavior occurs in the limit $\zeta \to 0$. When

ζ is large, the parabolic character prevails and the equation (2.25) can be handled directly.

The coefficients $C^\sharp$ which enter in the definition of $\mathcal{G}$ have limits at infinity in z, see (2.8). Therefore, $\mathcal{G}(z,p,\zeta)$ converges with an exponential rate to a limit $\mathcal{G}^\infty(p,\zeta)$ when z tends to $+\infty$.

LEMMA 2.5 (Spectral analysis of $\mathcal{G}$). *i) There are $c>0$ and $\rho_1>0$ such that for p in a compact neighborhood of $\underline{p}$, $|\zeta|\geq\rho_1$ with $\gamma\geq 0$, and $z\in[0,\infty[$, $\mathcal{G}(z,p,\zeta)$ has N eigenvalues, counted with their multiplicities, in $\operatorname{Re}\mu>0$ and N eigenvalues in $\operatorname{Re}\mu<0$. They satisfy $|\operatorname{Re}\mu|\geq c\langle\zeta\rangle$.*

ii) When $\zeta\neq 0$ and $\gamma\geq 0$, $\mathcal{G}^\infty(p,\zeta)$ has N eigenvalues, counted with their multiplicities, in $\operatorname{Re}\mu>0$ and N eigenvalues in $\operatorname{Re}\mu<0$.

iii) When $\zeta=0$, $\mathcal{G}^\infty(p,0)$ has 0 as a semi-simple eigenvalue, of multiplicity N. The nonvanishing eigenvalues are those of $(B^\infty_{d,d})^{-1}A^\infty_d$.

PROOF. **a)** When ζ is large, we use the quasi-homogeneity to write

$$\mathcal{M}=\langle\zeta\rangle^2\hat{M}+O(\langle\zeta\rangle)\,,\quad \mathcal{A}=\langle\zeta\rangle\hat{A}+O(1),$$

where

$$(2.31)\qquad \begin{cases}\hat{M}=(B^\sharp_{d,d})^{-1}\Big((i\hat{\tau}+\hat{\gamma})\mathrm{Id}+\displaystyle\sum_{j,k\geq 1}^{d-1}\hat{\eta}_j\hat{\eta}_k B^\sharp_{j,k}\Big)\\[2ex] \hat{A}=-i\displaystyle\sum_{k=1}^{d-1}\hat{\eta}_k(B^\sharp_{d,d})^{-1}(B^\sharp_{k,d}+B^\sharp_{d,k})\end{cases}$$

with

$$\hat{\tau}=\frac{\tau}{\langle\zeta\rangle^2}\,,\quad \hat{\gamma}=\frac{\gamma}{\langle\zeta\rangle^2}\,,\quad \hat{\eta}=\frac{\eta}{\langle\zeta\rangle}.$$

Thus

$$\begin{pmatrix}\langle\zeta\rangle\mathrm{Id} & 0\\ 0 & \mathrm{Id}\end{pmatrix}\mathcal{G}\begin{pmatrix}\langle\zeta\rangle^{-1}\mathrm{Id} & 0\\ 0 & \mathrm{Id}\end{pmatrix}=\langle\zeta\rangle\begin{pmatrix}0 & \mathrm{Id}\\ \hat{M} & \hat{A}\end{pmatrix}+O(1).$$

Tracing back the definitions, $\hat{\mu}$ is an eigenvalue of the matrix $\hat{\mathcal{G}}$, coefficient of the $\langle\zeta\rangle$ term in the right hand side, if and only if $-(i\hat{\tau}+\hat{\gamma})$ is an eigenvalue of

$$\sum_{j,k=1}^{d}\xi_j\xi_k B_{j,k}(b,w_p(z))$$

with $\xi_d=-i\mu$ and $(\xi_1,\dots,\xi_{d-1})=\hat{\eta}$. If μ belongs to imaginary axis, ξ_d is real and by (H1) one must have $\hat{\gamma}\leq -c|\xi|^2$. For $\hat{\gamma}\geq 0$, this implies that $\xi=0$, and therefore that $\hat{\tau}-i\hat{\gamma}=0$, which contradicts that $\langle\hat{\zeta}\rangle=1$. Thus $\hat{\mathcal{G}}$ has no eigenvalues on the imaginary axis. Therefore, the number of eigenvalues in $\operatorname{Re}\mu>0$ and in $\operatorname{Re}\mu<0$ is independent of $\hat{\zeta}$ when $\hat{\gamma}\geq 0$. Moreover, when $\hat{\eta}=0$, $\hat{\mathcal{G}}$ reduces to

$$\begin{pmatrix}0 & \mathrm{Id}\\ (B^\sharp_{d,d})^{-1} & 0\end{pmatrix}.$$

The eigenvalues are the square roots of the eigenvalues of $B^{-1}_{1,1}$, and therefore N of them are in $\operatorname{Re}\mu>0$ and N in $\operatorname{Re}\mu<0$.

By a standard perturbation argument, for $\langle\zeta\rangle$ large, the eigenvalues of $\mathcal{G}$ are $\mu=\langle\zeta\rangle\hat{\mu}+O(1)$, where $\hat{\mu}$ is an eigenvalue of $\hat{\mathcal{G}}$, and $i)$ of the lemma follows.

b) Similarly, tracing back the definitions and using (2.7), μ is an eigenvalue of $\mathcal{G}^\infty(p,\zeta)$ if and only if $-\tau + i\gamma$ is an eigenvalue of

$$\sum_{j=1}^{d} \eta_j A_j(b, a+c) - i \sum_{j,k=1}^{d} \xi_j \xi_k B_{j,k}(b, a+c)$$

with $\xi_d = -i\mu$ and $(\xi_1, \dots, \xi_{d-1}) = \eta$. If $\operatorname{Re}\mu = 0$, ξ is real and (H3) implies that $\gamma \le -c(|\mu|^2 + |\eta|^2)$. For $\gamma \ge 0$, this implies that $\gamma = 0$, $\mu = 0$ and $\eta = 0$. Thus the matrix above vanishes, the eigenvalue $-\tau$ must be zero and therefore, $\zeta = 0$. This shows that $\mathcal{G}^\infty$ has no eigenvalues on the imaginary axis when $\zeta \ne 0$ and $\gamma \ge 0$

The number of eigenvalues in $\operatorname{Re}\mu > 0$ and in $\operatorname{Re}\mu < 0$ is independent of (p,ζ) when $\zeta \ne 0$ and $\gamma \ge 0$. Letting z tend to ∞, *i)* implies that for ζ large, N eigenvalues lie on each half plane.

c) When $\zeta = 0$, one has

$$\mathcal{G}^\infty(p, 0) = \begin{pmatrix} 0 & \mathrm{Id} \\ 0 & B_{d,d}^{-1} A_d(b, a+c) \end{pmatrix}.$$

By (H1) (H4) $B_{1,1}^{-1} A_1(b, a+c)$ is invertible when p remains in a neighborhood of $\underline{p}$. Thus the eigenvalues of $\mathcal{G}^\infty$ are zero with multiplicity N, and the eigenvalues of $B_{1,1}^{-1} A_1$. □

LEMMA 2.6 (Conjugation to constant coefficient). *For all $\underline{\zeta} \in \mathbb{R}^{d+1}$ with $\underline{\gamma} \ge 0$, there is a neighborhood ω of $(\underline{p}, \underline{\zeta})$ and there is a matrix $\mathcal{W}$ defined and C^∞ on $[0, \infty[\times \omega$ such that*

i) $\mathcal{W}^{-1}$ is uniformly bounded and there is $\theta > 0$ such that

$$|\mathcal{W}(z, p, \zeta) - \mathrm{Id}| \le C e^{-\theta z}, \tag{2.32}$$

ii) $\mathcal{W}$ satisfies

$$\partial_z \mathcal{W} = \mathcal{G}(z)\mathcal{W}(z) - \mathcal{W}(z)\mathcal{G}^\infty. \tag{2.33}$$

PROOF. Consider (2.33) as an ordinary (linear) differential equation in the space of matrices. Because $\mathcal{G}$ converges exponentially to $\mathcal{G}^\infty$, it has the form

$$\partial_z \mathcal{W} = \mathcal{L}\mathcal{W} + \delta G(z)\mathcal{W},$$

where $\mathcal{L}$ is the constant coefficient operator $\mathrm{ad}\mathcal{G}^\infty = [\mathcal{G}^\infty, \cdot\,]$, and $\delta G(z)$ is the left multiplication by $\mathcal{G}(z) - \mathcal{G}(\infty) = O(e^{-\delta z})$. Now we apply the Gap Lemma of [**GZ**], [**ZH**], [**Z**], which asserts that associated with the eigenvalue 0 and the eigenvector Id of $\mathcal{L}$, there is a solution of (2.33) satisfying (2.32). Recall that $\mathcal{W}$ is obtained as the solution of

$$\mathcal{W}(z) = \mathrm{Id} + \int_0^z e^{(z-s)\mathcal{L}} \Pi_-(s)\delta G(s)\mathcal{W}(s)ds - \int_z^\infty e^{(z-s)\mathcal{L}} \Pi_+ \delta G(s)\mathcal{W}(s)ds$$

where Π_+ [resp Π_-] is the spectral projector on the sum of the generalized eigenspaces of $\mathcal{L}$ associated with eigenvalues in $\operatorname{Re}\mu > -\kappa$ [resp. $\operatorname{Re}\mu < -\kappa$] where κ is chosen in $]0, \delta[$ such that $\mathcal{L}$ has no eigenvalues on $\{\operatorname{Re}\mu = \kappa\}$. Together with the estimates in [**GZ**], [**ZH**], [**Z**] which prove the existence of a solutions such that $\mathcal{W} - \mathrm{Id} = O(e^{-\theta z})$ for $\theta < \kappa$, this shows that one can choose $\mathcal{W}$ depending smoothly on the parameters, as long as the eigenvalues of $\mathcal{L}$, which are differences of eigenvalues of $\mathcal{G}^\infty$, remain separated by a line $\mathrm{Re}\,\mu = \kappa$ for some $\kappa \in]0, \delta[$. This is true locally.

Consider $D(z) := \det \mathcal{W}(z)$. Then

$$\partial_z D(z) = \mathrm{tr}(\mathcal{G}(z) - \mathcal{G}^\infty)\, D(z)\,. \tag{2.34}$$

This clearly implies that $D(z)$ never vanishes on $[0, \infty[$. In addition, since $D(z) = 1 + O(e^{-\theta z})$, this also provides uniform bounds for $D(z)$ and $1/D(z)$. To prove (2.34), denote by $(W_1, \dots, W_{2N})$ [resp denote by $(G_1, \dots, G_{2N})$] the columns of $\mathcal{W}$ [resp. $\mathcal{G}$]. Then (2.33) implies that

$$\begin{aligned}\partial_z D = \sum_j \det\big[W_1, \dots, \mathcal{G}(z) W_j, \dots W_{2N}\big] \\ - \sum_j \det\big[W_1, \dots, \mathcal{W} G_j(\infty), \dots W_{2N}\big]\,.\end{aligned}$$

Next use the following algebraic identities for matrices $\mathcal{W}$ and $\mathcal{G}$ with columns $(W_1, \dots, W_{2N})$ and $(G_1, \dots, G_{2N})$:

$$\begin{aligned}\sum_j \det\big[W_1, \dots, \mathcal{G} W_j, \dots W_{2N}\big] = (\mathrm{tr}\,\mathcal{G}) \det \mathcal{W}\,, \\ \sum_j \det\big[W_1, \dots, \mathcal{W} G_j, \dots W_{2N}\big] = (\mathrm{tr}\,\mathcal{G}) \det \mathcal{W}\,,\end{aligned}$$

which are quite clear when $(W_1, \dots, W_{2N})$ is a basis, and extend algebraically to general $\mathcal{W}$. □

The substitution $U = \mathcal{W} U_1$ transforms the equation (2.25) into (2.30) with $F_1 = \mathcal{W}^{-1} F$ and $\Gamma_1(p, \zeta) = \Gamma \mathcal{W}^{-1}(0, p, \zeta) v$. In particular, let $\mathbb{E}_-(p, \zeta)$ [resp. $\mathbb{F}_-(p, \zeta)$] denote the space of initial data $U(0)$ [resp. $U_1(0)$] such that the corresponding solution of $\partial_z U = \mathcal{G}(z, p, \zeta) U$ [resp. $\partial_z U_1 = \mathcal{G}^\infty(p, \zeta) U_1$] is bounded as z tends to infinity. Then

$$\mathbb{E}_-(p, \zeta) = \mathcal{W}(0, p, \zeta) \mathbb{F}_-(p, \zeta)\,. \tag{2.35}$$

Corollary 2.7. *$\mathbb{E}_-(p, \zeta)$ and $\mathbb{F}_-(p, \zeta)$ have dimension N and vary smoothly with (p, ζ) when $\zeta \neq 0$ and $\gamma \geq 0$. In addition,*

$$\mathbb{F}_-(p, \zeta) \cap \ker \Gamma_1(p, \zeta) = \{0\}\,. \tag{2.36}$$

Proof. Since $\mathbb{F}_-$ is the spectral subspace for $\mathcal{G}^\infty$ associated to eigenvalues lying in $\mathrm{Re}\,\mu < 0$, it has dimension N by Lemma 2.5 and varies smoothly with the parameters (p, ζ) when $\zeta \neq 0$. The identity (2.36) follows from Assumption 1.4 which means that $\mathbb{E}_-(p, \zeta) \cap \ker \Gamma_1(p, \zeta) = \{0\}$. □

Remark 2.8. Lemmas 2.5, 2.6 and the first statement in Corollary 2.7 only use Assumptions 1.1 and 1.2. The transversality (2.36) relies on the additional Assumption 1.4.

2.4. Kreiss analysis. To prove estimates for solutions of the constant coefficient equation (2.30), we follow Kreiss' analysis, with a slight but important generalization of the block structure condition, needed to treat general viscosities in the nonstrictly hyperbolic case. (In the simpler, strictly hyperbolic or Laplacian viscosity case, the usual block structure condition will suffice; see Remark 2.11b.) When $\zeta \neq 0$ the block diagonalization of $\mathcal{G}^\infty$ is quite easy since the eigenvalues remain away from the imaginary axis. When working near $\underline{\zeta} = 0$, one has to push further the analysis of $\mathcal{G}^\infty(p, \zeta)$. We proceed in two steps.

LEMMA 2.9. *There is a C^∞ invertible matrix $\mathcal{V}$ defined on a neighborhood ω_0 of $(\underline{p}, 0)$ such that $\mathcal{V}^{-1}\mathcal{G}^\infty\mathcal{V}$ has the block diagonal form*

$$\mathcal{V}^{-1}\mathcal{G}^\infty\mathcal{V} = \begin{pmatrix} H & 0 \\ 0 & P \end{pmatrix} := \mathcal{G}_2 \tag{2.37}$$

with $H(p, 0) = 0$ and $P(p, 0) = (B^\infty_{d,d})^{-1}A^\infty_d$ and

$$\mathcal{V}(p, 0) = \begin{pmatrix} \mathrm{Id} & (A^\infty_d)^{-1}B^\infty_{d,d} \\ 0 & \mathrm{Id} \end{pmatrix} \tag{2.38}$$

The eigenvalues of P satisfy $|\mathrm{Re}\,\mu| \geq c$ for some $c > 0$ and

$$H = -(A^\infty_d)^{-1}\big((i\tau + \gamma)\mathrm{Id} + \sum_{j=1}^{d-1} i\eta_j A^\infty_j\big) + O(|\zeta|^2). \tag{2.39}$$

PROOF. By (H1) (H4), 0 is not an eigenvalue of $(B^\infty_{1,1})^{-1}A^\infty_1$. Lemma 2.5 implies that, on a small neighborhood Ω_0 of the origin, there is a smooth family of matrices $\mathcal{V}_1$ as indicated in the lemma.

The form of H follows from a direct perturbation argument. Moreover, if μ is an eigenvalue of $P(p, 0)$, then 0 is an eigenvalue of $\mu A^\infty_d - \mu^2 B^\infty_{d,d}$. Thus (H2) implies that $\mathrm{Re}\,\mu \neq 0$. This remains true for ζ in a neighborhood of 0. □

Next, we analyze the structure of H. The block reduction of the first order part is the key part in Kreiss' analysis. We show that, with suitable modification, the analysis can be extended to H. Introduce polar coordinates for ζ:

$$\begin{aligned} &\zeta = \rho\check{\zeta} = \rho(\check{\tau}, \check{\gamma}, \check{\eta}), \quad \text{with} \quad \rho = |\zeta|, \quad |\check{\zeta}| = 1, \\ &H(p, \zeta) = \rho\check{H}(p, \check{\zeta}, \rho). \end{aligned} \tag{2.40}$$

LEMMA 2.10 (The generalized block structure condition). *For all $\underline{\check{\zeta}}$ with $\underline{\check{\gamma}} \geq 0$ there is a neighborhood $\check{\omega}$ of $(\underline{p}, \underline{\check{\zeta}}, 0)$ in $\mathbb{R}^{d+1} \times S^d \times \mathbb{R}$ and there are matrices $V(p, \check{\zeta}, \rho)$ C^∞ on $\check{\omega}$ such that $V^{-1}\check{H}V$ has the following block diagonal structure:*

$$V^{-1}\check{H}V = \mathcal{Q}(p, \check{\zeta}) + \rho\mathcal{R}(p, \check{\zeta}, \rho), \tag{2.41}$$

with

$$\mathcal{Q} = \begin{bmatrix} \mathcal{Q}_1 & \cdots & 0 \\ \vdots & \ddots & \vdots \\ 0 & \cdots & \mathcal{Q}_{\underline{k}} \end{bmatrix}, \qquad \mathcal{R} = \begin{bmatrix} \mathcal{R}_1 & \cdots & 0 \\ \vdots & \ddots & \vdots \\ 0 & \cdots & \mathcal{R}_{\underline{k}} \end{bmatrix} \tag{2.42}$$

and

$$(2.43)\qquad \mathcal{Q}_k = \begin{bmatrix} Q_k & \cdots & 0 \\ \vdots & \ddots & \vdots \\ 0 & \cdots & Q_k \end{bmatrix}, \qquad \mathcal{R}_k = \begin{bmatrix} R^k_{1,1} & \cdots & R^k_{1,\alpha_k} \\ \vdots & \ddots & \vdots \\ R^k_{\alpha_k,1} & \cdots & R^k_{\alpha_k,\alpha_k} \end{bmatrix},$$

where the subblocks Q_k, $R^k_{p,q}$ *are* $\nu_k \times \nu_k$ *matrices, and the blocks* $\mathcal{Q}_k$, $\mathcal{R}_k$ *are* $\alpha_k\nu_k \times \alpha_k\nu_k$ *matrices. Moreover,* $\mathcal{Q}_k$, $\mathcal{R}_k$ *satisfy one of the conditions i) or ii) below if* $\underline{\check\gamma} > 0$, *and one of the conditions i) to iv) when* $\underline{\check\gamma} = 0$*:*

i) the spectrum of $Q_k(p, \check\zeta)$ *is contained in the open half plane* $\{\mathrm{Re}\,\mu > 0\}$*;*

ii) the spectrum of $Q_k(p, \check\zeta)$ *is contained in* $\{\mathrm{Re}\,\mu < 0\}$*.*

iii) $\nu_k = 1$, Q_k *is purely imaginary when* $\check\gamma = 0$, *and* $\partial_{\check\gamma}(\mathrm{Re}\,Q_k)\mathrm{Re}\,\mathcal{R}_k$ *is positive definite, where* $\mathrm{Re}\,\mathcal{R}_k := \frac{1}{2}(\mathcal{R}_k + \mathcal{R}_k^*)$.

iv) $\nu_k > 1$, Q_k *has purely imaginary coefficients when* $\check\gamma = 0$, *there is* $\mu_k \in \mathbb{R}$ *such that*

$$(2.44)\qquad Q_k(\underline{p}, \underline{\check\zeta}) = i \begin{bmatrix} \mu_k & 1 & 0 & \\ 0 & \mu_k & \ddots & 0 \\ & \ddots & \ddots & 1 \\ & & \cdots & \mu_k \end{bmatrix},$$

$$(2.45)\qquad R^k_{p,q}(\underline{p}, \underline{\check\zeta}, 0) = \begin{bmatrix} * & 0\dots0 \\ \vdots & 0\dots0 \\ r^k_{p,q} & 0\dots0 \end{bmatrix},$$

and $(\partial_{\check\gamma}\mathrm{Re}\,a_k)\mathrm{Re}\,R^\flat_k$ *is positive definite at* $(\underline{p}, \underline{\check\zeta}, 0)$, *where* a_k *is the lower left hand corner of* Q_k *and* $R^\flat_k$ *is the* $\alpha_k \times \alpha_k$ *matrix with entries* $r^k_{p,q}$.

REMARKS 2.11. **a)** This result (more precisely, assertions (2.41)–(2.44)) was originally established in [**Z**], by a closely related argument; see Observations 4.11–4.12 and equations (4.102)–(4.103) of that reference. In the main analysis of [**Z**] there appears an additional "foliated structure" hypothesis (H6) under which dependence on η is effectively suppressed in the bifurcation problem (2.41), and this was used in an essential way in establishing the $L^1 \to L^p$ estimates on which the analysis of [**Z**] is based. The fact that we may dispense with any such auxiliary hypotheses here reflects both the correctness of the L^2 norm for the problem at hand, and the power/generality of the Kreiss symmetrizer construction.

(There appears also in the main analysis of [**Z**] the assumption that $A(\eta, \xi)$ and $B(\eta, \xi)$ be simultaneously symmetrizable, under which the matrices $R^\flat_k$ may be chosen to be diagonal, recovering a "weak block structure" in (2.43). However, the latter is only a technical convenience, and could be removed.)

b) When $\rho = 0$, the block structure reduction $V^{-1}\check{H}V = \mathcal{Q}$ with $\mathcal{Q}$ as in (2.42) is established in [**Mé3**]. The construction there extends to the case where the eigenvalues of the entire symbol $iA(\eta, \xi) + \rho B(\eta, \xi)$ are of constant multiplicity with respect to *all* parameters including ρ. This holds in particular, when $A(\eta, \xi)$ is strictly hyperbolic or when $B(\eta, \xi) = |\eta|^2 + |\xi|^2$ corresponds to "artificial" Laplacian viscosity. However, constant multiplicity *with respect to* ρ typically fails for multiple characteristics. Indeed, a necessary condition is that, for all $\xi \in \mathbb{R}^d$,

$\Pi \sum_{j,k=1}^{d} B^{jk}\xi_j\xi_k\Pi$ be a scalar multiple of the identity, where Π is the eigenprojection associated with the multiple eigenvalue of $\sum_{j=1}^{d} A_j\xi_j$; this is violated, for example, for the "effective" strictly parabolic systems associated with gas dynamics and MHD (see [**HoZ**] for a description of effective viscosity and its relation with time-asymptotic behavior).

The independent proof of Lemma 2.10 is postponed to Appendix A. We just give now a flavor of the arguments, assuming constant multiplicity with respect to all parameters as in Remark b) above. One first performs a block reduction of $\check{H}$, isolating the purely imaginary eigenvalues from the eigenvalues with non vanishing real part of $\check{H}(\underline{p}, \check{\underline{\zeta}}, 0)$. The later case is easily handled. The former case occurs only when $\check{\underline{\gamma}} = 0$. In this case, consider a purely imaginary eigenvalue $\underline{\mu} = i\check{\underline{\xi}}_d$ of $\check{H}(\underline{p}, \check{\underline{\zeta}}, 0)$. Note that $\check{\underline{\xi}} = (\check{\underline{\eta}}, \check{\underline{\xi}}_d) \neq 0$ since $(\check{\underline{\tau}}, \check{\underline{\eta}}) \neq 0$. Suppose that $\lambda(p, \xi, \rho)$ an eigenvalue of constant multiplicity of $\sum \xi_j A_j^\infty(p) - i\rho \sum_{j,k} \xi_j\xi_k B_{j,k}^\infty(p)$ such that $\check{\underline{\tau}} + \lambda(\underline{p}, \check{\underline{\xi}}, 0) = 0$.

The key point is the following remark. By (H2), $\lambda(p, \xi, \rho)$ is real when $\rho = 0$ and ξ is real. Moreover, it satisfies

$$\operatorname{Im} \lambda(\underline{\xi}, 0) = 0 \,, \quad \operatorname{Im} \lambda(\underline{\xi}, \rho) \leq -c\rho \quad \text{for } \rho > 0 \text{ small}.$$

Therefore, $\partial_\rho \lambda(\underline{\xi}, 0) < 0$. This implies that the eigenvalue equation

$$\mathcal{F}(\check{\tau}, \check{\gamma}, \check{\xi}, \rho) := \check{\tau} - i\check{\gamma} + \lambda(\check{\xi}, \rho) = 0$$

satisfies

$$\partial_{\check{\gamma}} \operatorname{Im} \mathcal{F}(\underline{p}, \check{\underline{\zeta}}, 0) < 0 \quad \text{and} \quad \partial_\rho \operatorname{Im} \mathcal{F}(\underline{p}, \check{\underline{\zeta}}, 0) < 0 \,. \tag{2.46}$$

In particular, by the implicit function theorem, if $\check{\underline{\xi}}_d$ is a simple root (hyperbolic mode), there is a unique semi-simple eigenvalue $\mu(p, \check{\tau}, \check{\gamma}, \check{\eta}, \rho)$ of multiplicity α_k equal to the multiplicity of λ, close to $\underline{\mu}$. In this case, $\mathcal{Q}_k + \rho\mathcal{R}_k = \mu \mathrm{Id}$. Moreover, (2.46) implies that $\partial_{\check{\gamma}} \operatorname{Re} \mu$ and $\partial_\rho \operatorname{Re} \mu$ do not vanish and have the same sign. This is why the sign condition occurs in blocks *iii*) which correspond to semi-simple real purely imaginary eigenvalues of $H(\underline{p}, \check{\underline{\zeta}}, 0)$.

If $\check{\underline{\xi}}_d$ is a multiple root of the eigenvalue equation $\mathcal{F} = 0$, we repeat the proof in [**Mé3**]. The proof of the block reduction is unchanged. In this case, the matrices $\mathcal{R}_k$ are also block-diagonal, meaning that $R_{p,q}^k = 0$ when $p \neq q$. The integer ν_k is the order of the root $\check{\underline{\xi}}_d$ and α_k the multiplicity of λ. Moreover, all the matrices $Q_p + \rho R_{p,p}^k$, $p \in \{1, \dots, \alpha_k\}$, are equal. That they are purely imaginary when $\check{\gamma} = \rho = 0$ follows from Ralston's Lemma (see [**Ral**], [**Ch-P**]). The property of the lower left hand corner of this matrix follows from (2.46), exactly as the condition $\partial_{\check{\gamma}} \operatorname{Re} a \neq 0$ in the hyperbolic case ([**Ch-P**]). We also have $\partial_\rho \operatorname{Re} a \neq 0$ and the two derivatives have the same sign.

2.5. Construction of symmetrizers. Next we construct symmetrizers for (2.30). To warm up, we consider first the easy case where $\underline{\zeta} \neq 0$.

LEMMA 2.12 (Symmetrizers for medium frequencies). *For all* $\underline{\zeta} \neq 0$ *with* $\underline{\gamma} \geq 0$ *there is a neighborhood* ω *of* $(\underline{p}, \underline{\zeta})$ *and there is a* C^∞ *matrix* $\mathcal{S}(p, \zeta)$ *on* Ω *such that*

$$\mathcal{S} = \mathcal{S}^*, \tag{2.47}$$

$$\operatorname{Re}(\mathcal{S}\mathcal{G}^\infty) \geq \operatorname{Id}, \tag{2.48}$$

$$\mathcal{S} \geq \operatorname{Id} - C(\Gamma_1)^*\Gamma_1. \tag{2.49}$$

where C *is independent of* $(p, \zeta) \in \omega$.

PROOF. By Lemma 2.5, the eigenvalues of $\mathcal{G}^\infty(\underline{p}, \underline{\zeta})$ are away from the imaginary axis. Hence, there is a smooth invertible matrix $\mathcal{V}$ on a neighborhood ω of $(\underline{p}, \underline{\zeta})$, such that

$$\mathcal{V}^{-1}\mathcal{G}^\infty\mathcal{V} = \begin{pmatrix} G_+ & 0 \\ 0 & G_- \end{pmatrix}$$

where $G_\pm$ have their spectrum in $\pm\operatorname{Re}\mu > 0$. In the terminology of boundary value problems this is an elliptic situation. The symmetrizer is built in the form

$$\mathcal{S} = (\mathcal{V}^{-1})^* \begin{pmatrix} \kappa_+ S_+ & 0 \\ 0 & -\kappa_- S_- \end{pmatrix} \mathcal{V}^{-1}$$

with $S_\pm$ symmetric, positive definite, such that

$$\operatorname{Re}(S_+G_+) \geq \operatorname{Id}, \quad -\operatorname{Re}(S_-G_-) \geq \operatorname{Id}.$$

For instance, one can choose

$$S_+ = 2\int_0^\infty e^{-tG_+^*}e^{-tG_+}\,dt$$

and use a similar expression for S_-. If κ_+ and κ_- are large enough, (2.48) holds. Moreover, the form of $\mathcal{S}$ implies that there are constants c and C such that

$$(\mathcal{S}V, V) \geq c\kappa_+|\Pi_+V|^2 - C\kappa_-|\Pi_-V|^2. \tag{2.50}$$

where Π_+ [resp. Π_-] is the spectral projector of $\mathcal{G}^\infty$ on the space generated by generalized eigenvectors associated to eigenvalues in $\operatorname{Re}\mu > 0$ [resp. $\operatorname{Re}\mu < 0$]. These projectors are smooth functions of (p, ζ) in a neighborhood of $(\underline{p}, \underline{\zeta})$ since the two groups of eigenvalues remain separated. The Evans-Lopatinski stability condition (2.36) implies that $\ker\Pi_+ \cap \ker\Gamma_1 = \{0\}$. Moreover each space has dimension equal to N. Thus there is a constant C such that for (p, ζ) in a neighborhood of $(\underline{p}, \underline{\zeta})$, one has:

$$\forall V \in \mathbb{C}^{2N}: \qquad |V| \leq C\big(|\Pi_+(p, \zeta)V| + |\Gamma_1(p, \zeta)V|\big).$$

Thus, for κ_+/κ_- large enough, the right hand side of (2.50) is larger than $c'\kappa_+|V|^2 - C|\Gamma_1 V|^2$ and (2.49) follows for κ_+ large enough. □

For $\underline{\zeta} = 0$, we construct a symmetrizer for the matrix $\mathcal{G}_2 = \mathcal{V}_1^{-1}\mathcal{G}^\infty\mathcal{V}_1$ in (2.37), and more precisely for each block P and H. The associated boundary conditions are $\Gamma_2 = \Gamma_1\mathcal{V} = \Gamma\mathcal{W}(0, \cdot)\mathcal{V}$. We will use the notations of Lemma 2.10. In particular, we use the polar coordinates (2.40), S^d denoting the sphere $|\check{\zeta}| = 1$ and S^d_+ the closed half sphere where $\check{\gamma} \geq 0$.

LEMMA 2.13 (Symmetrizers for low frequencies). *i) There is a neighborhood* ω *of* $(\underline{p}, 0)$ *and there is a* C^∞ $N \times N$ *matrix* S_2 *on* ω *such that*

$$S_2 = (S_2)^*, \quad \operatorname{Re}(S_2P) \geq \operatorname{Id}.$$

ii) There is a C^∞ matrix $\check S_1$ on a neighborhood $\check\omega$ of $\{\underline{p}\} \times S^d_+ \times \{0\}$ such that $\check S_1 = (\check S_1)^*$ and

$$\operatorname{Re}(\check S_1 \check H) = \sum (\check V_l)^* \check K_l \check V_l \tag{2.51}$$

where $\{\check V_l\}$ is a finite collection of C^∞ $N \times N$ matrices and $\{\check K_l\}$ is a finite collection of C^∞ matrices having the following block structure

$$\check K_l = \begin{bmatrix} \check B_1 & \cdots & 0 \\ \vdots & \ddots & \vdots \\ 0 & \cdots & \check B_q \end{bmatrix} \tag{2.52}$$

with either $\check B_j = \check B_j^*$ positive definite or $\check B_j = \check\gamma \check B_{j,0} + \rho \check B_{j,1}$ with $\check B_{j,0}$ and $\check B_{j,0}$ positive definite. Moreover, $\sum \check V_l^* \check V_l$ is positive definite on $\check\omega$.

iii) The matrix $\mathrm{S} = \begin{bmatrix} \check S_1 & 0 \\ 0 & \check S_2 \end{bmatrix}$ with

$$\check S_2(p, \check\zeta, \rho) = \check S_2(p, \rho\check\zeta) \tag{2.53}$$

satisfies

$$(\mathrm{S}U, U) + C|\Gamma_2 U|^2 \geq |U|^2 \tag{2.54}$$

for some constant C independent of the parameters in $\check\omega$.

The proof is given in Appendix A. It is a modification of Kreiss' construction where the new ingredient is to control the dependence on the additional parameter ρ.

We now consider the case where ζ is large. The reduction to (2.30) is not true uniformly and we make a direct analysis of (2.25). This analysis is possible in the high frequency regime, because the parabolic properties are dominant. As in the proof of Lemma (2.5), we introduce "parabolic polar coordinates at infinity"

$$\begin{aligned} &\hat\zeta = (\hat\tau, \hat\eta, \hat\gamma) = (\lambda^2\tau, \lambda\eta, \lambda^2\gamma) \quad \text{with} \\ &\lambda = \langle\zeta\rangle^{-1} = (\tau^2 + \gamma^2 + |\eta|^4)^{\frac14}. \end{aligned} \tag{2.55}$$

and λ is small. Then

$$\mathcal{M}(z, p, \zeta) = \langle\zeta\rangle^2 \hat{\mathcal{M}}(z, p, \hat\zeta, \lambda) \quad \mathcal{A}(z, p, \zeta) = \langle\zeta\rangle \hat{\mathcal{A}}(z, p, \hat\zeta, \lambda)$$

where $\hat{\mathcal{M}}$ and $\hat{\mathcal{A}}$ are smooth for p close to $\underline{p}$, $\hat\zeta$ in the "sphere" $\hat S^d := \{\langle\zeta\rangle = 1\}$ and $\lambda \in [-1, 1]$. We denote by $\hat S^d_+$ the closed half sphere $\{\hat\gamma \geq 0\}$. It is convenient to reduce $\mathcal{G}$ to first order as in the proof of Lemma 2.5, introducing the change of unknowns

$$u_1 = \langle\zeta\rangle u, \quad v_1 = v. \tag{2.56}$$

Then, (2.25) is transformed into

$$\partial_z U_1 = \langle\zeta\rangle \mathcal{G}_1(z) U_1 + F_1\,, \quad \Gamma U(0) = u(0) = 0\,, \tag{2.57}$$

$$\mathcal{G}_1(z, p, \zeta) = \hat{\mathcal{G}}_1(z, p, \hat\zeta, \lambda) := \begin{pmatrix} 0 & \mathrm{Id} \\ \hat{\mathcal{M}} & \hat{\mathcal{A}} \end{pmatrix}.$$

LEMMA 2.14 (symmetrizers for high frequencies). *There is a neighborhood* $\hat{\omega}$ *of* $\{\underline{p}\} \times \hat{S}^d_+ \times \{0\}$ *and there is a* C^∞ *self adjoint matrix* $\hat{\mathcal{S}}$ *on* $[0, +\infty[\times\hat{\omega}$ *such that*

i) $\hat{\mathcal{S}}$ *and its derivatives converges with an exponential rate at* $z = +\infty$.

ii) $\mathrm{Re}\,(\hat{\mathcal{S}}\hat{\mathcal{G}}_1) \geq \mathrm{Id}$.

iii) $\hat{\mathcal{S}}_{|z=0} \geq \mathrm{Id} - C\Gamma^*\Gamma$.

PROOF. We note that the coefficients $C^\sharp(z,p)$ of $\hat{\mathcal{M}}$ and $\hat{\mathcal{A}}$ are functions of (z,p) through the substitution $C^\sharp(z,p) = C(p, w(p,z))$. Thus $\hat{\mathcal{G}}_1$ can be written as

$$\hat{\mathcal{G}}_1(z, p, \hat{\zeta}, \lambda) = \widetilde{\mathcal{G}}_1(p, w(p,z), \hat{\zeta}, \lambda).$$

Moreover, $w(p,\cdot)$ takes its values in a compact set. Therefore, we construct $\hat{\mathcal{S}}$ as a function of

$$\hat{\mathcal{S}}(z, p, \hat{\zeta}, \lambda) = \widetilde{\mathcal{S}}(p, w(z,p), \hat{\zeta}, \lambda)$$

Lemma 2.5 implies that for λ small the spectrum of $\widetilde{\mathcal{G}}_1$ remains in a compact set which does not intersect the imaginary axis when $\hat{\gamma} \geq 0$. Therefore, the spectral projectors $\Pi_\pm(p, w, \hat{\zeta}, \lambda)$ on the invariant spaces $\mathbb{F}^\infty_\pm(p, w, \hat{\zeta}, \lambda)$ of $\widetilde{\mathcal{G}}_1(p, w, \hat{\zeta}, \lambda)$ associated to eigenvalues with positive/negative real part are defined and smooth for λ small. There is a smooth matrix $\hat{\mathcal{V}}(p, w, \hat{\zeta}, \lambda)$ such that

$$\hat{\mathcal{V}}^{-1}\widetilde{\mathcal{G}}_1\hat{\mathcal{V}} = \begin{pmatrix} P_+ & 0 \\ 0 & P_- \end{pmatrix} \tag{2.58}$$

where the spectrum of P_+ [resp. P_-] remains in a compact subset of $\{\mathrm{Re}\,\mu > 0\}$ [resp. $\{\mathrm{Re}\,\mu < 0\}$]. One constructs $\widetilde{t\mathcal{S}}$ as in Lemma 2.12

$$\widetilde{\mathcal{S}} = (\mathcal{V}^{-1})^* \begin{pmatrix} \kappa_+ S_+ & 0 \\ 0 & -\kappa_- S_- \end{pmatrix} \mathcal{V}^{-1}.$$

with $\kappa_\pm > 0$ large enough. This implies properties $i)$ and $ii)$.

To get the property $iii)$, we use that there is C and λ_0 small, such that for (p, w) in a neighborhood of $(\underline{p}, 0)$, $\hat{\zeta} \in \hat{S}^d_+$ and $\lambda \in [0, \lambda_0]$

$$|U| \leq C(|\Gamma U| + |\Pi_+(p, w, \hat{\zeta}, \lambda)U|).$$

This follows from the transversality

$$\mathbb{F}^\infty_-(p, w, \hat{\zeta}, \lambda) \cap \ker \Gamma = \{0\}. \tag{2.59}$$

Indeed, for $z = 0$, we note that $w(p, 0) = c$ is small and therefore one can chose the parameters $\kappa_\pm$ such that $\widetilde{\mathcal{S}}(p, w, \hat{\zeta}, \lambda) \geq \mathrm{Id} - C\Gamma^*\Gamma$ provided that w and λ are small, implying $iii)$.

We now prove (2.59) when $\lambda = 0$ and $w = 0$. The property remains true when w and λ are small enough. Towards this end, we give a link between the spaces $\mathbb{F}^\infty_-(p, 0, \hat{\zeta}, 0)$ and the spaces $\widetilde{\mathbb{E}}_-(b, u, \zeta)$ introduced in section 1.

Recall that $\widetilde{\mathbb{E}}_-(b, u, \zeta)$ is the set of $(\dot{u}, \Lambda^{-1}\dot{v})$ such that the solution $U = (u, v)$ of the homogeneous equation (2.25) with initial data $(\dot{u}, \dot{v})$ is bounded. Introduce then the change of unknowns and variables

$$u_2(\hat{z}) = u(\hat{z}/\Lambda), \qquad v_2(\hat{z}) = \frac{1}{\Lambda}v(\hat{z}/\Lambda).$$

Then the homogeneous equation (2.25) is transformed into

$$\partial_{\hat{z}} U_2 = \mathcal{G}_2(\hat{z}/\Lambda, p, \hat{\zeta}, \lambda)U_2 \tag{2.60}$$

with

$$\mathcal{G}_2(z,p,\hat\zeta,\lambda) = \begin{pmatrix} 0 & \mathrm{Id} \\ \frac{\langle\hat\zeta\rangle^2}{\Lambda^2}\hat{\mathcal{M}} & \frac{\langle\hat\zeta\rangle}{\Lambda}\hat{\mathcal{A}} \end{pmatrix}$$

The coefficients of $\hat{\mathcal{M}}$ and $\hat{\mathcal{A}}$ are functions of p and $w(p,z)$. Remember that $w(p,0) = 0$. Therefore, as λ tends to zero (or equivalently, as Λ tends to infinity),

$$\mathcal{G}_2(\hat z/\Lambda, p, \hat\zeta, \lambda) \to \widetilde{\mathcal{G}}_1(p,0,\hat\zeta,0) = \begin{pmatrix} 0 & \mathrm{Id} \\ \hat M_0 & \hat A_0 \end{pmatrix}$$

where $\hat M_0$ and $\hat A_0$ are the evaluation at $w = 0$ of the functions $\hat M$ and $\hat A$ defined at (2.31). By definition, $\widetilde{\mathbb{E}}_-(p,u,\zeta)$ is the set of initial data $(\dot u_1, \dot v_2)$ such that the corresponding solution of (2.60) is bounded. The similar space for the equation

$$\partial_z U = \widetilde{\mathcal{G}}_1(p,0,\hat\zeta,0) \tag{2.61}$$

is $\mathbb{F}^\infty_-(p,0,\hat\zeta,0)$. Using the exponential convergence of the coefficients and the fact that $\widetilde{\mathcal{G}}_1$ has no eigenvalues on the imaginary axis, one easily shows that

$$\widetilde{\mathbb{E}}_-(p,u,\zeta) \to \mathbb{F}^\infty_-(p,0,\hat\zeta,0)$$

as $\lambda \to 0$ and $\zeta = (\lambda^{-2}\hat\tau, \lambda^{-1}\hat\eta, \lambda^{-2}\hat\gamma)$ with $\hat\zeta \neq 0$, $\hat\gamma \geq 0$.

The uniform stability condition implies that the spaces $\widetilde{\mathbb{E}}_-$ and $\ker\Gamma$ are uniformly transverse to each other. This remains true for the limit, implying (2.59). □

REMARK 2.15. The transversality condition (2.59) is equivalent to the requirement that the problem

$$-\partial_z^2 u + \hat A_0 \partial_z u + \hat M_0 u = 0\,, \qquad u(0) = 0 \tag{2.62}$$

has no solution in $H^2([0,\infty[)$. Suppose that the parabolic problem is symmetric, i.e. that there is a matrix $S(b,u)$ such that

$$\mathrm{Re} \sum \xi_j \xi_k S(b,u) B_{j,k}(b,u)$$

is definite positive. If $u \in H^2$ satisfies (2.62), taking the real part of the scalar product in L^2 with $(SB_{d,d})u$ yields

$$\begin{aligned} \mathrm{Re}\,\big(SB_{d,d}u,u\big) &+ \sum_{j=1}^{d-1} \mathrm{Re}\, i\eta_j \big((B_{j,d} + B_{d,j})\partial_z u, u\big) \\ &+ \sum_{j,k=1}^{d-1} \mathrm{Re}\,\eta_j\eta_k \big((B_{j,k}u,u\big) + \gamma\big(Su,u\big) = 0\,. \end{aligned}$$

The assumption on B implies that the sesquilinear form on the left hand side is coercive on the space $H^1_0([0,+\infty[)$, as easily seen by extending u by 0 for negative z. Thus (2.62) has no non trivial H^2. Therefore, the symmetry of the Parabolic operator implies the transversality (2.59) and thus that the uniform stability condition is automatically satisfied for large ζ.

2.6. Proof of Theorem 2.3. We prove the estimate (2.29) in the three different regimes.

a) *Medium frequencies.* Lemmas 2.12 and 2.2 imply that for all $\underline{\zeta} \neq 0$, there is a neighborhood ω of $(\underline{u}, \underline{\zeta})$ such that the solutions of (2.30) satisfy

$$\|U_1\|^2 + |U_1(0)|^2 \leq \|F_1\|^2.$$

Therefore, shrinking ω if necessary, the solutions $U = \mathcal{W}U_1$ of (2.25) satisfy

$$\|U\|^2 + |U(0)|^2 \leq \|F\|^2,$$

which implies (2.29) for $(p, \zeta) \in \omega$.

b) *Low frequencies.* Let $\mathcal{V}$ be given by Lemma 2.9. Then $U_2 = \mathcal{V}^{-1}U_1$ satisfies

$$\partial_z U_2 - \mathcal{G}_2 U_2 = F_2 := \mathcal{V}^{-1}F_1\,, \qquad \Gamma_2 V_2(0) = 0\,. \tag{2.63}$$

We use the symmetrizers S_1 and S_2 given by Lemma 2.13. One has

$$\operatorname{Re}(S_1 H) \geq c\rho(\check{\gamma} + \rho) = c(\gamma + |\zeta|^2) \approx h^2\,, \quad \operatorname{Re}(S_2 P) \geq \operatorname{Id}.$$

Therefore, the components (u_2, v_2) and (f_2, g_2) of U_2 and F_2 respectively, satisfy

$$h^2\|u_2\|^2 + \big(S_1 u_2(0), u_2(0)\big) \leq C\frac{1}{h^2}\|f_2\|^2\,,$$

$$\|v_2\|^2 + \big(S_2 v_2(0), v_2(0)\big) \leq C\|g_2\|^2\,.$$

Summing, and using the third part of Lemma 2.13, we obtain that

$$h^2\|u_2\|^2 + \|v_2\|^2 + |U_2(0)| \leq C\Big(\frac{1}{h^2}\|f_2\|^2 + \|g_2\|^2\Big)\,;$$

thus

$$h^2\|u_2\| + h\|v_2\| + h|U_2(0)| \leq C\big(\|f_2\| + h\|g_2\|\big) \leq C\|F_2\|\,.$$

Lemma (2.9) implies that $U_1 = \mathcal{V}U_2$ and $F_1 = \mathcal{V}F_2$ satisfy

$$u_1 = O(1)U_2\,, \quad v_1 = O(1)v_2 + O(\zeta)u_2\,, \quad F_2 = O(1)F_1\,.$$

Since $h \leq 1$ and $|\zeta| \leq h$, this implies that $|\zeta|h \leq h^2$ and

$$h^2\|u_1\| + h\|v_1\| + h|U_1(0)| \leq C\|F_1\|\,.$$

Since $\mathcal{W} = \operatorname{Id} + O(e^{-\theta z})$, $U = \mathcal{W}U_1$, one has

$$u = O(1)U_1\,, \quad v = O(1)v_1 + O(e^{-\theta z})u_1\,, \quad F_1 = O(1)F\,.$$

We use here the following inequality:

$$\|e^{-\theta z}u_1\|_{L^2([0,\infty[)} \leq C\left(|u_1(0)| + \|\partial_z u_1\|_{L^2([0,\infty[)}\right). \tag{2.64}$$

Here, the equation (2.30) implies that $\partial_z u_1 = v_1$. Thus

$$\|e^{-\theta z}u_1\| \leq C\big(|U_1(0)| + \|v_1\|\big),$$

and therefore

$$h^2\|u\| + h\|v\| + h|U(0)| \leq C\|F\|\,.$$

This implies that (2.29) holds for (p, ζ) in a small neighborhood of $(\underline{p}, 0)$.

c) *High frequencies.* With notations as in Lemma 2.14, we use the symmetrizer

$$\mathcal{S}(z, p, \zeta) = \hat{\mathcal{S}}(z, p, \hat{\zeta}, \lambda).$$

Then

$$\operatorname{Re}\mathcal{S}\mathcal{G}_1 \geq \operatorname{Id}$$

and Lemma 2.2 implies that

$$\langle\zeta\rangle\|U_1\|^2 + |U_1(0)|^2 \le C\Big(\frac{1}{\langle\zeta\rangle}\|F_1\|^2 + \|\partial_z\mathcal{S}\|_{L^\infty}\|U_1\|^2\Big).$$

Thus, for $\langle\zeta\rangle$ large enough, the second term in the right can be absorbed by the left hand side and

$$\langle\zeta\rangle\|U_1\| + \langle\zeta\rangle^{1/2}|U_1(0)| \le C\|F_1\|\,.$$

Going back to $u = \langle\zeta\rangle^{-1}u_1$ and $v = v_1$ this shows that (2.29) is valid for ζ large enough. □

3. Pieces of paradifferential calculus

We construct symmetrizers as para-differential operators in the variables (t, y), depending on the parameters x, ε and γ. The analysis in section 2 shows that different homogeneities occur in different regions of the frequency space. We first recall from [**Mok**], [**Mé1**] (see also [**Mé2**]) the handbook of results about the homogeneous paradifferential calculus with parameter, to be used in the hyperbolic regime. It is a modification of the original calculus introduced by Bony ([**Bo**], [**Mey**]). Next we will state the similar results for a semi-classical calculus associated to symbols having a parabolic type homogeneity. This calculus will be used in the middle and high frequency regimes. The symmetrizers will provide the basic L^2 estimates. However, the estimates of the derivatives also require a microlocal analysis. Indeed, a direct differentiation of the equation reveals unbounded terms in ε^{-1} and a detailed analysis of commutators yields to distinguish the different regimes. This yields to study the action of the para-differential operators in conormal spaces $\mathcal{H}^m$ (see the definition (1.28)). The idea is to replace the base space L^2 by $\mathcal{H}^m$, and to show that the calculus extends to this framework. Since we make strong assumptions on the smoothness of the coefficients, this follows from repeated application of Leibniz formula.

3.1. The homogeneous calculus. We consider operators on $\mathbb{R}^d$. The variables are denoted $\tilde{y} = (t, y)$ and the frequency variables $\tilde{\eta} = (\tau, \eta)$. The symbols and operators also depend on a parameter γ which plays a distinguished role. We denote by $\mathbb{R}^{d+1}_+$ the set of frequencies $\zeta := (\tilde{\eta}, \gamma) \in \mathbb{R}^{d+1}\setminus\{0\}$ such that $\gamma \ge 0$ and by S^d_+ the set of $(\tilde{\eta}, \gamma) \in \mathbb{R}^{d+1}_+$ such that $|\zeta| = 1$.

DEFINITION 3.1 (Symbols). Let $\mu \in \mathbb{R}$. i) Γ^μ_0 denotes the space of locally L^∞ functions $a(\tilde{y}, \zeta)$ on $\mathbb{R}^d \times \mathbb{R}^{d+1}_+$ which are C^∞ with respect to ζ and such that for all $\alpha \in \mathbb{N}^d$ there is a constant C_α such that

$$\forall(\tilde{y}, \zeta), \quad |\partial^\alpha_{\tilde{\eta}} a(\tilde{y}, \zeta)| \le C_\alpha\,|\zeta|^{\mu-|\alpha|}\,. \tag{3.1}$$

ii) Γ^μ_1 denotes the space of symbols $a \in \Gamma^\mu_0$ such that for all j, $\partial_{\tilde{y}_j}a \in \Gamma^\mu_0$.

For example, functions $a(\tilde{y}, \zeta)$ which are C^∞ and homogeneous of degree m in $(\tilde{\eta}, \gamma) \in \mathbb{R}^{d+1}_+$ and bounded on $\mathbb{R}^d \times S^d_+$, are symbols in Γ^m_0.

In the applications, we consider families of symbols $a^\varepsilon(x)$ in Γ^m_k, depending on parameters $\varepsilon \in [0, 1[$ and $x \in [0, \infty[$. The key argument is that they are bounded in Γ^m_k. Moreover, we want to study the action of the operators in conormal spaces. Consider the following set of vector fields on $\mathbb{R}^{d+1}_+$ with variables $(t, y) \in \mathbb{R}^d$ and $x \ge 0$:

$$Z_0 = \partial_t\,, \quad Z_j = \partial_{y_j} \ \text{ for } 1 \le j \le d-1\,, \quad Z_d = x\partial_x\,. \tag{3.2}$$

They commute, and for $\alpha \in \mathbb{Z}^{d+1}$ $Z^\alpha = Z_0^{\alpha_0} \cdots Z_d^{\alpha_d}$.

DEFINITION 3.2. For $\mu \in \mathbb{R}$, $m \in \mathbb{N}$ and $k \in \{0,1\}$, $\Gamma^\mu_{k,m}$ is the set of functions $a(x, \tilde{y}, \zeta)$ such that for all $\alpha \in \mathbb{Z}^{d+1}$ with $|\alpha| \le m$, the functions $(Z^\alpha a)(x, \cdot, \cdot)$ form a bounded family in Γ^μ_k.

The spaces $\Gamma^\mu_{k,m}$ are equipped with semi-norms

$$\|a\|_{(\mu,k,m,N)} := \sup_{|\alpha|\le N} \sup_{|\beta|\le k} \sup_{|\sigma|\le m} \sup_{(x,\tilde{y},\zeta)} |\zeta|^{|\alpha|-m} \left| Z^\sigma \partial_\zeta^\alpha \partial_{\tilde{y}}^\beta a(x, \tilde{y}, \zeta) \right| . \tag{3.3}$$

A family of symbols is bounded in $\Gamma^\mu_{k,m}$ when for all N, the semi norms are bounded.

EXAMPLE 3.3. Suppose that $\mathrm{a}(x, \tilde{y}, \zeta, \rho)$ is a function on $[0,\infty[\times \mathbb{R}^d \times \mathbb{R}^{d+1}_+ \times [0,\infty[$, C^∞ with respect to (ζ, ρ), homogeneous of degree μ in ζ, supported in $\{\rho \le 1\}$ and such that $Z^\alpha \mathrm{a}$ and $Z^\alpha \partial_{\tilde{y}} \mathrm{a}$ are bounded on $[0,\infty[\times \mathbb{R}^d \times S^d_+ \times [0,1[$ for all $|\alpha| \le m$. Then, the family of symbols a^ε defined by

$$a^\varepsilon(x, \tilde{y}, \tilde{\eta}, \gamma) = \mathrm{a}\big(x, \tilde{y}, \zeta, \varepsilon|\zeta|\big) \tag{3.4}$$

is bounded in $\Gamma^\mu_{1,m}$.

The para-differential calculus is a quantization of symbols in $a \in \Gamma^\mu_0$ to which are associated operators denoted by T^γ_a; see Appendix B. They act in the scale of Sobolev spaces $H^s(\mathbb{R}^d)$. These spaces are equipped with the family of norms

$$\big| u \big|_{0,s,\gamma} := \Big(\int_{\mathbb{R}^d} (\gamma^2 + |\tilde{\eta}|^2)^s \, |\hat{u}(\tilde{\eta})|^2 \, d\tilde{\eta} \Big)^{\frac{1}{2}} . \tag{3.5}$$

Adding the normal variable x, we introduce the norms

$$\begin{aligned} \|u\|_{0,s,\gamma} &= \Big(\int \big| u(x,\cdot) \big|^2_{0,s,\gamma} \, dx \Big)^{\frac{1}{2}} , \\ \|u\|_{m,s,\gamma} &= \sum_{|\alpha|\le m} \gamma^{m-|\alpha|} \, \|Z^\alpha u\|_{0,s,\gamma} , \end{aligned} \tag{3.6}$$

which are parameter dependent norms on spaces called $\mathcal{H}^{0,s}$ and $\mathcal{H}^{m,s}$ respectively.

When $a \in \Gamma^\mu_{0,m}$, for all x, $a(x) = a(x,\cdot) \in \Gamma^\mu_0$, and the action of para-differential operators is extended to x-dependent functions:

$$(T^\gamma_a u)(x, \cdot) = T^\gamma_{a(x,\cdot)} u(x, \cdot). \tag{3.7}$$

The paradifferential calculus in $\mathbb{R}^d$, was introduced by J.M.Bony [**Bo**] (see also [**Mey**] , [**Hör**], [**Tay**]) with γ fixed, say $\gamma = 1$. The parameter dependent version T^γ_a is introduced in [**Mél**] [**Mok**] and applies in the scale of spaces $\mathcal{H}^{0,s}$. The extension to the scale $\mathcal{H}^{m,s}$ is immediate since one can construct the T^γ so that

$$Z T^\gamma_a u = T^\gamma_a Z u + T^\gamma_{Za} u . \tag{3.8}$$

This is explained in Appendix B, together with the details of the following results.

PROPOSITION 3.4 (Action). *i) When $a(\zeta)$ is a symbol independent of $\tilde{y}$, the operator T^γ_a is defined by the action of the Fourier multiplier a.*

ii) For all $a \in \Gamma^\mu_{0,m}$, the family of operators $\{T^\gamma_a\}_{\gamma\ge 1}$ is of order $\le \mu$, meaning that for $\gamma \ge 1$:

$$\|T^\gamma_a u\|_{m,s,\gamma} \le C \; \|u\|_{m,s+\mu,\gamma}$$

where C is independent of $\gamma \ge 1$ and u.

PROPOSITION 3.5 (Symbolic calculus). *Consider* $a \in \Gamma^{\mu}_{1,m}$ *and* $b \in \Gamma^{\mu'}_{1,m}$. *Then* $ab \in \Gamma^{\mu+\mu'}_{1,m}$ *and* $\{T^\gamma_a \circ T^\gamma_b - T^\gamma_{ab}\}_{\gamma\geq 1}$ *is of order* $\leq \mu+\mu'-1$, *meaning that for* $\gamma \geq 1$:

$$\|(T^\gamma_a T^\gamma_b - T^\gamma_{ab})u\|_{m,s,\gamma} \leq C\ \|u\|_{m,s+\mu+\mu'-1,\gamma}$$

where C *is independent of* $\gamma \geq 1$ *and* u.

If b *is independent of* $\tilde{y}$, *then* $T^\gamma_a \circ T^\gamma_b = T^\gamma_{ab}$.

These results extend to matrix valued symbols and operators.

PROPOSITION 3.6 (Adjoints). *Consider a matrix valued symbol* $a \in \Gamma^{\mu}_{1,0}$. *Denote by* $(T^\gamma_a)^*$ *the adjoint operator of* T^γ_a *in* $L^2(\mathbb{R}^{d+1}_+)$ *and by* $a^*(\tilde{y},\zeta)$ *the adjoint of the matrix* $a(\tilde{y},\zeta)$. *Then* $\{(T^\gamma_a)^* - T^\gamma_{a^*}\}_{\gamma\geq 1}$ *is of order* $\leq m-1$, *meaning that for* $\gamma \geq 1$:

$$\|((T^\gamma_a)^* - T^\gamma_{a^*})u\|_{0,s,\gamma} \leq C\|\nabla_{\tilde{y}} a\|_{(\mu,0,0,N)}\ \|u\|_{0,s+\mu-1,\gamma}$$

where C *and* N *only depend on the indices* s *and* μ.

PROPOSITION 3.7 (Gårding's inequality). *Consider a* $N\times N$ *matrix symbol* $a \in \Gamma^{\mu}_{1,0}$, *a* $N \times M$ *matrix symbol* $w \in \Gamma^0_{1,0}$, *and a scalar symbol* $\chi \in \Gamma^0_{1,0}$. *Assume that* $\chi w = w$ *and there is constant* $c > 0$ *such that*

$$\forall(x,\tilde{y},\zeta):\quad \chi^2(x,\tilde{y},\zeta)\,\mathrm{Re}\, a(x,\tilde{y},\zeta) \geq c\,\chi^2(x,\tilde{y},\zeta)\,|\zeta|^{\mu} \tag{3.9}$$

Then, there are C *that for all* $\gamma \geq 1$ *and* u

$$\frac{c}{2}\ \|T^\gamma_w u\|^2_{0,\frac{\mu}{2},\gamma} \leq \mathrm{Re}\,\big(T^\gamma_a T^\gamma_w u, T^\gamma_w u\big)_{L^2}\ +\ C\,\|u\|^2_{0,\frac{\mu}{2}-1,\gamma}$$

where C *is bounded the symbols remain bounded and* (3.9) *holds with a uniform constant* c.

REMARK 3.8. The meaning of the assumption is that $\mathrm{Re}\, a$ is definite positive on the support of w. The symbol χ only plays an intermediate role. We give two applications. Consider two open sets ω and $\widetilde{\omega}$ with ω relatively compact in $\widetilde{\omega}$. Consider two cones in $\mathbb{R}^{d+1}_+$, C and $\widetilde{\mathrm{C}}$ such that the base of the cone C, $\overline{\mathrm{C}} \cap S^d_+$ is relatively compact in $\mathrm{C} \cap S^d_+$.

1. Suppose that a and w are homogeneous of degree μ and 0 respectively, that w is supported in $\omega \times \mathrm{C}$ and that $\mathrm{Re}\, a \geq c|\zeta|^{\mu}$ on $\widetilde{\omega}\times\widetilde{\mathrm{C}}$. Then there is $\chi \in \Gamma^0_{1,0}$ supported in $\widetilde{\omega}\times\widetilde{\mathrm{C}}$ such that $\chi = 1$ on the support of w. Thus $\chi w = w$ and (3.9) holds.

2. Suppose that a^{ε} and w^{ε} are bounded families of homogeneous symbols of degree μ and 0 respectively with w supported in $\omega \times (\mathrm{C} \cap \mathrm{B}_{r/\varepsilon})$ where B_R denotes the ball $\{|\zeta| \leq R\}$. Suppose that $\mathrm{Re}\, a \geq c|\zeta|^{\mu}$ on $\widetilde{\omega} \times (\widetilde{\mathrm{C}} \cap \mathrm{B}_{r'/\varepsilon})$ where $r' > r$. Then, there is $\chi \in \Gamma^0_{1,0}$ supported in $\widetilde{\omega}\times(\widetilde{\mathrm{C}}\cap\mathrm{B}_{r'/\varepsilon})$ such that $\chi = 1$ on the support of w. Thus $\chi w = w$ and (3.9) holds

Bounded functions of $\tilde{y}$ are particular examples of symbols in the class Γ^0_0, independent of the frequency variables ζ. In this case, T^γ_a is called a para-product in [**Bo**]. In our case, we introduce the spaces $\mathcal{W}^{m,k}$ of function on $\mathbb{R}^{d+1}_+$ such that $Z^\alpha\partial^\beta_{\tilde{y}} u \in L^\infty(\mathbb{R}^{d+1}_+)$ for all $|\alpha| \leq m$ and $|\beta| \leq k$, equipped with the norms

$$\|a\|_{\mathcal{W}^{m,k}} = \sum_{|\alpha|\leq m}\sum_{|\beta|\leq k} \|Z^\alpha\partial^\beta_{\tilde{y}} u\|_{L^\infty(\mathbb{R}^{d+1}_+)}\,. \tag{3.10}$$

When $k = 0$, we simply denote by $\mathcal{W}^m$ the corresponding space.

PROPOSITION 3.9 (Para-products). *There is a constant C such that for all $a \in \mathcal{W}^{m,1}$ and all $u \in \mathcal{H}^{m,0}$*

$$\begin{aligned} \|au - T_a^\gamma u\|_{m,1,\gamma} &\leq C\,\|a\|_{\mathcal{W}^{m,1}}\|u\|_{m,0,\gamma}\,, \\ \gamma\|au - T_a^\gamma u\|_{m,0,\gamma} &\leq C\,\|a\|_{\mathcal{W}^{m,1}}\|u\|_{m,0,\gamma}\,, \\ \|a\partial_j u - T_a^\gamma \partial_j u\|_{m,0,\gamma} &\leq C\,\|a\|_{\mathcal{W}^{m,1}}\|u\|_{m,0,\gamma}\,. \end{aligned}$$

3.2. The semi-classical parabolic calculus. In the high frequency regime, the parabolic character of the equations prevails. In this regime, the quasi-homogeneity is associated to the dilations $\lambda \cdot (t, y) = (\lambda^2 t, \lambda y)$ and similarly $\lambda \cdot (\tau, \gamma, \eta) = (\lambda^2\tau, \lambda^2\gamma, \lambda\eta)$. The corresponding quasi-norm is

$$\langle\zeta\rangle = \left(\gamma^2 + \tau^2 + |\eta|^4\right)^{\frac{1}{4}}. \tag{3.11}$$

We also introduce the weight

$$\Lambda(\zeta) = (1 + \langle\zeta\rangle^4)^{\frac{1}{4}}. \tag{3.12}$$

Typical example of symbols are smooth quasi-homogeneous functions of degree μ away from the origin. They satisfy

$$|\partial_\zeta^\alpha a(\zeta)| \leq C_\alpha \langle\zeta\rangle^{m-\langle\alpha\rangle}$$

where, for $\alpha = (\alpha_\tau, \alpha_\eta) \in \mathbb{N} \times \mathbb{N}^{d-1}$:

$$\langle(\alpha_\tau, \alpha_\eta)\rangle := 2|\alpha_\tau| + |\alpha_\eta|.$$

Next we consider a semi-classical quantification of the symbols. In particular, when $a(\zeta)$ is independent of $\tilde{y}$, the associated operator is defined by the Fourier multiplier $a(\varepsilon\tilde{\eta}, \varepsilon\gamma)$:

$$P_a^{\varepsilon,\gamma} = a(\varepsilon D_{\tilde{y}}, \varepsilon\gamma). \tag{3.13}$$

Note that we use here the standard multiplication by ε, not the parabolic dilation $\varepsilon \cdot \tilde{\eta}$. We also extend the calculus to x dependent symbols and functions, x being considered as a parameter. We also consider the action in conormal spaces.

DEFINITION 3.10 (Symbols). Let $\mu \in \mathbb{R}$.

i) $\mathrm{P}\Gamma_0^\mu$ denotes the space of locally L^∞ functions $a(\tilde{y}, \zeta)$ on $\mathbb{R}^d \times \mathbb{R}_+^{d+1}$ which are C^∞ with respect to ζ and such that for all $\alpha \in \mathbb{N}^d$ there is a constant C_α such that

$$\forall(\tilde{y}, \zeta)\,, \quad |\partial_{\tilde{\eta}}^\alpha a(\tilde{y}, \tilde{\eta}, \gamma)| \leq C_\alpha\, \Lambda(\zeta)^{\mu-\langle\alpha\rangle}\,. \tag{3.14}$$

ii) $\mathrm{P}\Gamma_1^\mu$ denotes the space of symbols $a \in P\Gamma_0^\mu$ such that for all j, $\partial_{\tilde{y}_j} a \in P\Gamma_0^\mu$.

iii) For $m \in \mathbb{N}$ and $k \in \{0, 1\}$, $\mathrm{P}\Gamma_{k,m}^\mu$ is the set of functions $a(x, \tilde{y}, \zeta)$ such that for all $\alpha \in \mathbb{Z}^{d+1}$ with $|\alpha| \leq m$, the functions $(Z^\alpha a)(x, \cdot, \cdot)$ form a bounded family in $\mathrm{P}\Gamma_k^\mu$.

EXAMPLES 3.11. 1. If $a(\tilde{y}, \zeta)$ has compact support in ζ, is C^∞ in ζ and the derivatives $\partial_\zeta^\alpha a$ are bounded, then $a \in \mathrm{P}\Gamma_0^0$.

2. If $a(\tilde{y}, \zeta)$ is C^∞ and quasi-homogeneous of degree μ in ζ, with derivatives $\partial_\zeta^\alpha a$ bounded and bounded on $\mathbb{R}^d \times S_+^d$, then $(1 - \chi(\zeta)a) \in \Gamma_0^\mu$ for all $\chi \in C_0^\infty(\mathbb{R}^{d+1})$ which is equal to 1 on a neighborhood of the origin.

3. Suppose that $\mathrm{a}(x, \tilde{y}, \zeta, \delta)$ is a function on $[0, \infty[\times \mathbb{R}^d \times \mathbb{R}_+^{d+1} \times [0, \infty[$, C^∞ with respect to (ζ, ρ), quasi-homogeneous of degree μ in ζ, supported in $\{\delta \leq 1\}$

and such that for all $|\alpha| \leq m$, the derivatives $Z^\alpha \mathrm{a}$ and $Z^\alpha \partial_{\tilde{y}} \mathrm{a}$ are bounded on $[0, \infty[\times \mathbb{R}^d \times \{\langle \zeta \rangle = 1, \gamma \geq 0\} \times [0, 1[$. Then, the symbol defined by

$$a(x, \zeta) = (1 - \chi(\zeta))\, \mathrm{a}\big(x, \tilde{y}, \zeta, \frac{1}{\langle \zeta \rangle}\big) \tag{3.15}$$

belongs to $P\Gamma^\mu_{1,m}$ if $\chi \in C_0^\infty$ is equal to one near the origin.

The spaces $P\Gamma^\mu_{k,m}$ are equipped with semi-norms

$$\|a\|_{(\mu,k,m,N)} := \sup_{\langle \alpha \rangle \leq N} \sup_{|\beta| \leq k} \sup_{|\sigma| \leq m} \sup_{(x, \tilde{y}, \zeta)} \Lambda^{\langle \alpha \rangle - \mu} |Z^\sigma \partial_\zeta^\alpha \partial_{\tilde{y}}^\beta a(x, \tilde{y}, \zeta)|\,. \tag{3.16}$$

The natural scale of Sobolev spaces are the spaces PH^s of functions whose Fourier transform belong to the L^2 space with weight Λ^{2s}. Because we use a semi-classical analysis, this leads to introduce on PH^s the following family of norms

$$|u|_{0,s,\varepsilon,\gamma} := \Big(\int_{\mathbb{R}^d} \Lambda(\varepsilon \tilde{\eta}, \varepsilon \gamma)^{2s} \, |\hat{u}(\tilde{\eta})|^2 \, d\tilde{\eta} \Big)^{\frac{1}{2}} . \tag{3.17}$$

Adding the normal variable x, we introduce the norms

$$\begin{aligned} \|u\|_{0,s,\varepsilon,\gamma} &= \Big(\int |u(x, \cdot)|^2_{0,s,\varepsilon,\gamma} \, dx \Big)^{\frac{1}{2}} , \\ \|u\|_{m,s,\varepsilon,\gamma} &= \sum_{|\alpha| \leq m} \gamma^{m - |\alpha|} \|Z^\alpha u\|_{0,s,\varepsilon,\gamma} , \end{aligned} \tag{3.18}$$

which are parameter dependent norms on spaces called $P\mathcal{H}^{0,s}$ and $P\mathcal{H}^{m,s}$ respectively.

One first constructs the operators $P_a^{\varepsilon,\gamma}$ for symbols $a \in P\Gamma_0^\mu$, acting in the scale PH^s of functions of $\tilde{y}$. Next the action is extended to x dependent functions in $P\mathcal{H}^{0,s}$ and symbols in $P\Gamma^\mu_{0,0}$:

$$(P_a^{\varepsilon,\gamma} u)(x, \cdot) = P^{\varepsilon,\gamma}_{a(x,\cdot)} u(x, \cdot). \tag{3.19}$$

The study is finally extended to functions in $P\mathcal{H}^{m,s}$ and symbols in $P\Gamma^\mu_{0,m}$ using the identity

$$Z P_a^{\varepsilon,\gamma} u = P_a^{\varepsilon,\gamma} Z u + P_{Za}^{\varepsilon,\gamma} u\,. \tag{3.20}$$

We refer to Appendix B for details and proofs of the following results.

PROPOSITION 3.12 (Action). *i) When $a(\zeta)$ is a symbol independent of $\tilde{y}$, the operator $P_a^{\varepsilon,\gamma}$ is defined by the action of the Fourier multiplier $a(\varepsilon\tilde{\eta}, \varepsilon\gamma)$.*

ii) For all $a \in P\Gamma^\mu_{0,m}$ and $s \in \mathbb{R}$, there is C such that for $\varepsilon \in]0, 1]$, $\gamma \geq 1$ and $u \in P\mathcal{H}^{m,s}$:

$$\|P_a^{\varepsilon,\gamma} u\|_{m,s-\mu,\varepsilon,\gamma} \leq C\, \|u\|_{m,s,\varepsilon,\gamma}\,.$$

The constant C is bounded when a remains in a bounded set of $P\Gamma^\mu_{0,m}$.

iii) If $a \in P\Gamma_0^\mu$ is supported in $\mathbb{R}^d \times \{\Lambda(\zeta) \leq R\}$, then, for all u, the spectrum of $P_a^{\varepsilon,\gamma} u$ is contained in $\{\Lambda(\varepsilon\zeta) \leq 2R\}$

iv) There is $\delta > 0$, such that If $a \in P\Gamma_0^\mu$ is supported in $\mathbb{R}^d \times \{\Lambda(\zeta) \geq R\}$ then, for all u, the spectrum of $P_a^{\varepsilon,\gamma} u$ is contained in $\mathbb{R}^d \times \{\Lambda(\varepsilon\zeta) \geq \delta R\}$.

PROPOSITION 3.13 (Symbolic calculus). *Consider $a \in P\Gamma^\mu_{1,m}$ and $b \in P\Gamma^{\mu'}_{1,m}$. Then $ab \in P\Gamma^{\mu+\mu'}_{1,m}$ and there is C such that for $\varepsilon \in]0, 1]$, $\gamma \geq 1$ and $u \in P\mathcal{H}^{m,s}$:*

$$\|(P_a^{\varepsilon,\gamma} \circ P_b^{\varepsilon,\gamma} - P_{ab}^{\varepsilon,\gamma}) u\|_{m,s-\mu-\mu'+1,\varepsilon,\gamma} \leq C\, \varepsilon\, \|u\|_{m,s,\varepsilon,\gamma}\,.$$

The constant C is bounded when a and b remain bounded in $\mathrm{P}\Gamma^{\mu}_{1,m}$ and $\mathrm{P}\Gamma^{\mu'}_{1,m}$ respectively.

Moreover, if b is independent of $\tilde{y}$, then $P^{\varepsilon,\gamma}_a \circ P^{\varepsilon,\gamma}_b = P^{\varepsilon,\gamma}_{ab}$.

PROPOSITION 3.14 (Adjoints). *Consider a matrix valued symbol $a \in \mathrm{P}\Gamma^{\mu}_{1,0}$. Denote by $(P^{\varepsilon,\gamma}_a)^*$ the adjoint operator of $P^{\varepsilon,\gamma}_a$ in $L^2(\mathbb{R}^{d+1}_+)$ and by $a^*(x,\tilde{y},\zeta)$ the adjoint of the matrix $a(x,\tilde{y},\zeta)$. Then there is C such that for $\varepsilon \in]0,1]$, $\gamma \geq 1$ and $u \in P\mathcal{H}^{m,s}$:*

$$\|((P^{\varepsilon,\gamma}_a)^* - P^{\varepsilon,\gamma}_{a^*})u\|_{m,s-\mu+1,\varepsilon,\gamma} \leq C\,\varepsilon\,\|u\|_{m,s,\varepsilon,\gamma}.$$

The constant C is bounded when a remains in a bounded set of $\mathrm{P}\Gamma^{\mu}_{1,0}$.

PROPOSITION 3.15 (Gårding's inequality). *Consider a $N \times N$ matrix symbol $a \in \mathrm{P}\Gamma^{\mu}_{1,0}$, a $N \times M$ matrix symbol $w \in \mathrm{P}\Gamma^{0}_{1,0}$ and a scalar symbol $\chi \in \mathrm{P}\Gamma^{0}_{1,0}$. Assume that $\chi w = w$ and there is constant $c > 0$ such that*

$$\forall (x,\tilde{y},\zeta) : \quad \chi^2(x,\tilde{y},\zeta)\mathrm{Re}\, a(x,\tilde{y},\zeta) \geq c\chi^2(x,\tilde{y},\zeta)\,\Lambda(\zeta)^{\mu}. \tag{3.21}$$

Then, there is C such that for all $\gamma \geq 1$, all $\varepsilon \in]0,1]$ and u

$$\frac{c}{2}\,\|P^{\varepsilon,\gamma}_w u\|^2_{0,\frac{\mu}{2},\varepsilon,\gamma} \leq \mathrm{Re}\,\big(P^{\varepsilon,\gamma}_a P^{\varepsilon,\gamma}_w u, P^{\varepsilon,\gamma}_w u\big)_{L^2} + C\,\varepsilon^2\,\|u\|^2_{0,\frac{\mu}{2}-1,\varepsilon,\gamma}.$$

Moreover the constant C is bounded when the symbols remain in bounded sets and (3.21) is satisfied with a uniform constant c.

REMARK 3.16. Again, the meaning of the assumption is that $\mathrm{Re}\, a$ is definite positive on the support of w. We give here two examples. Consider two open sets ω and $\widetilde{\omega}$ with ω relatively compact in $\widetilde{\omega}$.

1. Consider two bounded open sets C and $\widetilde{\mathrm{C}}$ in $\mathbb{R}^{d+1}_+$ with C relatively compact in $\widetilde{\mathrm{C}}$. Suppose that w is supported in $\omega \times \mathrm{C}$ and that $\mathrm{Re}\, a \geq c|\zeta|^{\mu}$ on $\widetilde{\omega} \times \widetilde{\mathrm{C}}$. Then there is $\chi \in \mathrm{P}\Gamma^{0}_{1,0}$ supported in $\widetilde{\omega} \times \widetilde{\mathrm{C}}$ such that $\chi = 1$ on the support of w. Thus $\chi w = w$ and (3.21) holds.

2. Suppose that w is supported in $\omega \times \{\langle\zeta\rangle \geq R\}$ and that $\mathrm{Re}\, a \geq c\langle\zeta\rangle^{\mu}$ on $\widetilde{\omega} \times \{\langle\zeta\rangle \geq R'\}$ where $R > R' > 0$. Then, there is $\chi \in \Gamma^{0}_{1,0}$ supported in $\widetilde{\omega} \times \{\langle\zeta\rangle \geq R'\}$ such that $\chi = 1$ on $\omega \times \{\langle\zeta\rangle \geq R\}$. Thus $\chi w = w$ and (3.21) holds

Next we consider para-products, i.e. symbols independent of ζ in the class $\mathcal{W}^{m,1}$ introduced just before (3.10).

PROPOSITION 3.17 (Para-products). *For all $a \in \mathcal{W}^{m,1}(\mathbb{R}^d)$, there is a constant C such that for all $u \in P\mathcal{H}^{m,1}$, $\varepsilon \in]0,1]$, and $\gamma \geq 1$:*

$$\begin{aligned}\|au - P^{\varepsilon,\gamma}_a u\|_{m,1,\varepsilon,\gamma} + \sum_{|\alpha|=1} \varepsilon\|a\partial^{\alpha}_y u - P^{\varepsilon,\gamma}_a \partial^{\alpha}_y u\|_{m,0,\varepsilon,\gamma} \\ \leq C\varepsilon\|u\|_{m,0,\varepsilon,\gamma}\,,\end{aligned} \tag{3.22}$$

$$\begin{aligned}\gamma\|au - P^{\varepsilon,\gamma}_a u\|_{m,0,\varepsilon,\gamma} + \|a\partial_t u - P^{\varepsilon,\gamma}_a \partial_t u\|_{m,0,\varepsilon,\gamma} \\ + \sum_{|\alpha|=2} \varepsilon\|a\partial^{\alpha}_y u - P^{\varepsilon,\gamma}_a \partial^{\alpha}_y u\|_{m,0,\varepsilon,\gamma} \leq C\|u\|_{m,1,\varepsilon,\gamma}\,.\end{aligned} \tag{3.23}$$

COROLLARY 3.18. *For all $a \in \mathcal{W}^{m,2}(\mathbb{R}^d)$, there is a constant C such that for all $u \in P\mathcal{H}^{m,2}$, $\varepsilon \in]0,1]$, and $\gamma \geq 1$:*

$$\|au - P^{\varepsilon,\gamma}_a u\|_{m,2,\varepsilon,\gamma} \leq C\varepsilon\|u\|_{m,1,\varepsilon,\gamma}\,,$$

We now give a link between the two calculi. Remark first that for constant coefficient symbols a, T_a^γ and $P_a^{\varepsilon,\gamma}$ are Fourier multipliers by $a(\tilde\eta,\gamma)$ and $a(\varepsilon\tilde\eta,\varepsilon\gamma)$ respectively. Thus

$$P_a^{\varepsilon,\gamma} = T_{a^\varepsilon}^\gamma \quad \text{with} \quad a^\varepsilon(\zeta) = a(\varepsilon\zeta) \text{ when } a \text{ is independent of } (x,\tilde y)\,. \tag{3.24}$$

The next result extends this relation to symbols which also depend on the variables $\tilde y$. The proof is given in Appendix B.

PROPOSITION 3.19. *Suppose that $b \in \mathrm{P}\Gamma^0_{1,0}$ has compact support in ζ. Then the family of symbols*

$$\mathrm{b}^\varepsilon(x,\tilde y,\zeta) = b(x,\tilde y,\varepsilon\zeta) \tag{3.25}$$

is bounded in $\Gamma^0_{1,0}$ and there is a constant C such that for all u, $\varepsilon \in]0,1]$ and $\gamma \geq 1$:

$$\gamma\|T_{\mathrm{b}^\varepsilon}^\gamma u - P_b^{\varepsilon,\gamma}u\|_{L^2} + \|T_{\mathrm{b}^\varepsilon}^\gamma \nabla_{\tilde y} u - P_b^{\varepsilon,\gamma}\nabla_{\tilde y} u\|_{L^2} \leq C\|u\|_{L^2}\,. \tag{3.26}$$

Moreover, in the scale of norms (3.6)

$$\|T_{\mathrm{b}^\varepsilon}^\gamma u - P_b^{\varepsilon,\gamma}u\|_{0,0,\gamma} \leq C\|u\|_{0,-1,\gamma}\,. \tag{3.27}$$

3.3. Calculi on the boundary. We have developed above the para-differential calculus in the half space $\{x \geq 0\}$. By construction, x acts as a parameter, see the definitions (3.7), (3.19) and Appendix B. When $x = 0$, we obtain two calculi, still denoted by T^γ and $P^{\varepsilon,\gamma}$, on the boundary. We do not make specific statements in these case, they are in fact particular cases of the results above, provided that the set of vector fields $\{Z_j\}$ is restricted to the fields $\partial_{\tilde y_j}$. Taking traces, (3.7) and (3.19) imply that

$$\begin{cases} (T_a^\gamma u)_{|x=0} = T_{a_{|x=0}}^\gamma u_{|x=0}\,, \\ (P_a^{\varepsilon,\gamma} u)_{|x=0} = P_{a_{|x=0}}^{\varepsilon,\gamma} u_{|x=0}\,. \end{cases} \tag{3.28}$$

4. L^2 and conormal estimates near the boundary

In this section, we start the analysis of the stability of the linearized equations (1.25). We concentrate here on the most difficult part, which is the proof of the L^2 and $\mathcal{H}^m$ conormal estimates for functions localized near a point of the boundary. The main estimate is stated in Theorem 4.1. The strategy is as follows: we mimic the analysis of Section 2, substituting the paradifferential calculus to Fourier multipliers technics. We first replace the differential linearized equation by paradifferential equations, at the price of acceptable errors. This being done, we use the results of Section 3: the symbolic calculus only produces acceptable errors, and the calculus on symbols is precisely the calculus on Fourier multipliers developed in Section 2. Again we separate the three different regimes: low, medium and high frequencies. At the end, we glue the estimates together using a partition of unity. In each regime, the symmetrizers are paradifferential quantizations of the multipliers constructed in Section 2; these are used to prove the basic L^2 estimates. Finally, we prove the $\mathcal{H}^m$ estimates by differentiating the equations, which requires a lot of care due to the presence of singular terms in the equation.

4.1. The main estimate. Recall that $(Z_0, \ldots, Z_{\underline{k}})$ denote a basis of vector fields on $\mathbb{R}^{1+d}$ tangent to $\mathbb{R}\times\partial\Omega$. With $m \geq 0$, we consider u_0 in $W^{m+2,\infty}([-T_0, T_0]\times \Omega)$ and b_0 in $W^{m+2,\infty}([-T_0, T_0] \times \Omega)$. Following the notations of the introduction, b denotes the function $(t, x) \mapsto (t, x, b_0(t, x))$. We always assume that (b, u_0) is valued in a compact subset of $\mathcal{O}$.

In addition, we consider a family v^ε in $W^{m+2,\infty}([-T_0, T_0] \times \Omega)$ such that

$$\sup_{\varepsilon} \sup_{|J|\leq m} \|Z_J v^\varepsilon\|_{L^\infty} + \varepsilon\|\nabla_{t,x} Z_J v^\varepsilon\|_{L^\infty} + \varepsilon^2\|\nabla_x^2 Z_J v^\varepsilon\|_{L^\infty} < \infty . \tag{4.1}$$

where $Z_J = Z_{j_1} \cdots Z_{j-k}$ when $J = (j_1, \ldots, j_k)$ and $|J| = k$. Introduce

$$u_a^\varepsilon = W(b(t, x), u_0(t, x), \varphi(x)/\varepsilon) + \varepsilon v^\varepsilon(t, x) \tag{4.2}$$

The linearized equations from (1.1) around u_a^ε read

$$\begin{cases} \partial_t u + \displaystyle\sum_{j=1}^{d} A_j^\varepsilon \partial_j u - \varepsilon \sum_{j,k=1}^{d} B_{j,k}^\varepsilon \partial_{j,k}^2 u + \frac{1}{\varepsilon} E^\varepsilon u = f , \\ u_{|[-T_0,T_0]\times\partial\Omega} = 0 . \end{cases} \tag{4.3}$$

with

$$\begin{cases} A_j^\varepsilon v = \widetilde{A}_j v - \varepsilon \displaystyle\sum_k (\widetilde{B}'_{j,k} \cdot v)\partial_k u_a^\varepsilon - \varepsilon \sum_k (\widetilde{B}'_{k,j} \cdot \partial_k u_a^\varepsilon) v , \\ B_{j,k}^\varepsilon = \widetilde{B}_{j,k} , \\ E^\varepsilon v = \varepsilon \displaystyle\sum_j (\widetilde{A}'_j \cdot v)\partial_j u_a^\varepsilon - \varepsilon^2 \sum_{j,k} (\widetilde{B}'_{j,k} \cdot v)\partial_{j,k}^2 u_a^\varepsilon + \widetilde{B}''_{j,k}(v, \partial_j u_a^\varepsilon)\partial_k u_a^\varepsilon , \end{cases}$$

where $\widetilde{A}$ stands for the function A evaluated at $(b(t, x), u_a^\varepsilon(t, x))$ and A' is the derivative of A with respect to the variable u.

It is convenient to extend u_0, b and v^ε to $\mathbb{R} \times \Omega$, as $W^{m+2,\infty}$ functions. We still denote by u_0, b and v^ε the extended functions. We can assume that the extended v^ε is bounded in $W^{m+2,\infty}(\mathbb{R} \times \Omega)$ and that the extended functions are independent of t for $|t| \geq |T_1 > T_0$. We denote by u_a^ε the extension of u_a^ε given by (4.2) and we consider the linearized equation (4.3) around u_a^ε.

Consider a point $\underline{y} \in \partial\Omega$ and local coordinates $(y, x) \in \mathbb{R}^{d-1} \times \mathbb{R}$ near $\underline{y}$ such that $\underline{y}$ corresponds to 0 and the defining function φ of $\partial\Omega$ is x, so that Ω lies on the side $\{x > 0\}$. From now on, we work in these coordinates and restrict attention to functions u and f supported in a small fixed neighborhood $\widetilde{\omega}$ of $(\underline{t}, 0)$, where the coordinate patch is defined. We still denote by u, b, f etc the functions in the local coordinates. For convenience, we keep the notations in (4.3) for the linearized equations, x being the d-th spatial variable. However, the new coefficients A_j^ε and $B_{j,k}^\varepsilon$ involve the derivatives of the change of variables, which just introduce a new dependence of the coefficients on the function b.

We do not give the explicit relation between the coefficients in the new and old coordinates. We simply note that in the new set, they write as a principal term plus a remainder as follows

$$A_j^\varepsilon = \widetilde{A}_j^\sharp + \varepsilon A_{j,1}^\varepsilon , \quad B_{j,k}^\varepsilon = \widetilde{B}_{j,k}^\sharp , \quad E^\varepsilon = \widetilde{E}^\sharp + \varepsilon E_1^\varepsilon \tag{4.4}$$

where:

- $A_j^\sharp$, $B_{j,k}^\sharp$ and $E^\sharp$ are C^∞ functions of $z \in [0, \infty[$ and $p = (b, u, c)$ in a neighborhood of $\underline{p} = (\underline{b}, \underline{u}, 0)$. Here, $\underline{b} = b(\underline{t}, 0)$ and $\underline{u} = u_0(\underline{t}, 0)$. They have the general structure

$$F^\sharp(z, p) = F(b, W(b, u, z) + c).$$

Moreover, when $x = 0$, that is when $b \in \mathcal{B}_\partial$, the coefficients $A_j^\sharp$, $B_{j,k}^\sharp$ and $E^\sharp$ are those introduced in section two at (2.6).

- In (4.4), $\widetilde{F}^\sharp$ stands for the function $F^\sharp$ evaluated at $(p^\varepsilon(t, y, x), z)$ with

$$p^\varepsilon(t, y, x) = \big(b(t, y, x), u_0((t, y, x), \varepsilon v^\varepsilon(t, y, x)\big).$$

By (4.1) there holds

$$\sup_{\varepsilon \in]0,1]} \sup_{|J| \le m} \|Z_J \nabla_{t,y,x} p^\varepsilon\|_{L^\infty(\omega)} < +\infty \,. \tag{4.5}$$

- The remainders $A_{j,1}^\varepsilon$ and E_1^ε are smooth functions of (t, y, x), u_0^ε and its first and second derivatives, εv and its first order derivatives, $\varepsilon^2 v$ and its second order derivatives and $z = x/\varepsilon$. In particular, (4.1) implies that the remainders satisfy

$$\sup_{\varepsilon \in]0,1]} \sup_{|J| \le m} \|Z_J F_1^\varepsilon\|_{L^\infty(\omega)} < +\infty \,. \tag{4.6}$$

We write (4.3) in the condensed form

$$\begin{cases} -\varepsilon \partial_x^2 u + A^\varepsilon \partial_x u + \dfrac{1}{\varepsilon} M^\varepsilon u = (B_{d,d}^\varepsilon)^{-1} f \,, \\ u_{x=0} = 0 \,, \end{cases} \tag{4.7}$$

where

$$\begin{cases} A^\varepsilon = (B_{d,d}^\varepsilon)^{-1} \Big(A_d^\varepsilon - \displaystyle\sum_{j=1}^{d-1} (B_{j,d}^\varepsilon + B_{d,j}^\varepsilon) \varepsilon \partial_j \Big) \,, \\ M^\varepsilon = (B_{d,d}^\varepsilon)^{-1} \Big(\varepsilon \partial_t + \displaystyle\sum_{j=1}^{d-1} A_j^\varepsilon \varepsilon \partial_j - \sum_{j,k=1}^{d-1} B_{j,k}^\varepsilon \varepsilon^2 \partial_{j,k}^2 + E^\varepsilon \Big) \,. \end{cases}$$

A^ε is a first order operator in $\varepsilon \partial_y$, while M^ε is first order in $\varepsilon \partial_t$ and second order in $\varepsilon \partial_y$. Write (4.7) as a system for $U = {}^t(u, v)$, $v = \varepsilon \partial_x u$:

$$\begin{cases} \partial_x U - \dfrac{1}{\varepsilon} G^\varepsilon U = F \,, \\ \Gamma U_{|x=0} = 0 \,, \end{cases} \tag{4.8}$$

with

$$\Gamma \begin{pmatrix} u \\ v \end{pmatrix} = u \,, \quad G^\varepsilon = \begin{pmatrix} 0 & \mathrm{Id} \\ M^\varepsilon & A^\varepsilon \end{pmatrix} \,, \qquad F = \begin{pmatrix} 0 \\ -(B_{d,d}^\varepsilon)^{-1} f \end{pmatrix} \,.$$

To prove weighted estimates for the solutions of (4.3) or (4.8) we introduce $\widetilde{U} = e^{-\gamma t} U$ and $\widetilde{f} = e^{-\gamma t} f$. Then, (4.8) is equivalent to

$$\begin{cases} \partial_x \widetilde{U} - \dfrac{1}{\varepsilon} G^{\varepsilon,\gamma} \widetilde{U} = \widetilde{F} = \begin{pmatrix} 0 \\ -\widetilde{f} \end{pmatrix} \,, \\ \Gamma \widetilde{U}_{|x=0} = 0 \,, \end{cases} \tag{4.9}$$

where $G^\varepsilon = \begin{pmatrix} 0 & \mathrm{Id} \\ M^{\varepsilon,\gamma} & A^\varepsilon \end{pmatrix}$ and $M^{\varepsilon,\gamma}$ has the same definition as M^ε with ∂_t replaced by $\partial_t + \gamma$.

Following the analysis of section 2, we consider the following weight functions: with $\zeta := (\tau, \gamma, \eta)$,

$$\Lambda = \Lambda(\varepsilon\zeta) := \Big(1 + (\varepsilon\gamma)^2 + (\varepsilon\tau)^2 + |\varepsilon\eta|^4\Big)^{\frac{1}{4}}, \tag{4.10}$$

$$\varphi = \begin{cases} \big(\gamma + \varepsilon|\zeta|^2\big)^{\frac{1}{2}} & \text{when} \quad |\varepsilon\zeta| \le 1, \\ \approx \varepsilon^{-\frac{1}{2}} & \text{when} \quad 1 \le |\varepsilon\zeta| \le 2, \\ \frac{\Lambda}{\sqrt{\varepsilon}} \approx \big(\gamma + |\tau| + \varepsilon|\eta|^2\big)^{\frac{1}{2}} & \text{when} \quad |\varepsilon\zeta| \ge 2. \end{cases} \tag{4.11}$$

Note that the three terms have the same order when $|\varepsilon\zeta| \approx 1$.

Given a weight function $\psi(\tau, \eta)$, we introduce the norm

$$\big|u\big|_{(\psi)} = \Big(\int \psi(\tau,\eta)^2\, |\hat{u}(\tau,\eta)|^2\, d\tau d\eta\Big)^{\frac{1}{2}}, \tag{4.12}$$

where $\hat{u}$ is the Fourier transform of $u(t, y)$ defined on $\mathbb{R}^d$. When u also depends on the variable x, we denote by $\|u\|_{(\psi)}$ the norm

$$\|u\|_{(\psi)} = \Big(\int_0^\infty |u(x)|_{(\psi)} dx\Big)^{\frac{1}{2}}. \tag{4.13}$$

We use different weight functions, φ, φ^2, φ/Λ etc. In these case, the weights and the norms depend on the parameters ε and γ, but we do not make this dependence explicit and use the notations $\|\cdot\|_{(\varphi)}$ etc.

Next, we introduce the conormal spaces. In the local coordinates, we can choose

$$Z_0 = \partial_t\,, \quad Z_j = \partial_{y_j} \quad \text{for } 1 \le j \le d-1, \quad Z_d = x\partial_x\,. \tag{4.14}$$

Indeed, one should take $\ell(x)\partial_x$ in place of $x\partial_x$ with $\ell(x) = x$ for $|x| \le 1$ and $\ell(x) = 1$ for x large. Because, we only consider compactly supported functions, we make the choice above.

For $m \in \mathbb{N}$, define

$$\|u\|_{m,(\varphi)} = \sum_{|\alpha| \le m} \gamma^{m-|\alpha|} \|Z^\alpha u\|_{(\varphi)}. \tag{4.15}$$

They are norms, depending on the parameters ε and γ on a space called $\mathcal{H}^m_\varphi$. Note that the norms $\|\cdot\|_{m,s,\varepsilon,\gamma}$ introduced in section 3, correspond to the weights Λ^s. The L^2 norm corresponds to $m = 0$ with weight $\psi = 1$. We denote it by $\|\cdot\|_0 = \|\cdot\|_{0,(1)}$. More generally, we note $\|\cdot\|_m = \|\cdot\|_{m,(1)}$ the norms associated to the weight $\psi = 1$.

THEOREM 4.1. *There are a neighborhood ω of $(\underline{t}, 0)$ in $\mathbb{R}^{1+d}_+$, constants γ_0, $\varepsilon_0 > 0$ and C, such that for all $\widetilde{U}$ and $\widetilde{f}$ supported in ω and satisfying* (4.9), *for all $\gamma \ge \gamma_0$, $\varepsilon \in]0, \varepsilon_0]$ with $\varepsilon\gamma \le 1$, one has*

$$\|\widetilde{u}\|_{m,(\varphi^2)} + \frac{1}{\sqrt{\varepsilon}}\|\widetilde{v}\|_{m,(\varphi)} + |\widetilde{v}(0)|_{m,(\varphi/\sqrt{\Lambda})} \le C\|\widetilde{f}\|_m\,. \tag{4.16}$$

Dropping the tildes, we will prove the apparently weaker estimate

$$\begin{aligned} &\|u\|_{m,(\varphi^2)} + \frac{1}{\sqrt{\varepsilon}}\|v\|_{m,(\varphi)} + |v(0)|_{m,(\varphi/\sqrt{\Lambda})} \le \\ &\qquad C\Big(\|f\|_m + \|u\|_{m,(\Lambda)} + \|v\|_m + |v(0)|_m\Big). \end{aligned} \tag{4.17}$$

Since $\varphi \geq c\sqrt{\gamma}$ and $\varphi^2 \geq c\sqrt{\gamma}\Lambda$, one has

$$\|u\|_{m,(\Lambda)} \lesssim \gamma^{-1/2}\|u\|_{m,(\varphi^2)}, \quad \|v\|_m \lesssim \gamma^{-1/2}\|v\|_{m,(\varphi)}$$
$$|v(0)|_m \lesssim \gamma^{-1/4}|v(0)|_{m,(\varphi/\sqrt{\Lambda})}.$$

where $\lesssim$ means that the left hand side is estimated by constant times the right hand side, with a constant independent of ε, γ, U and f. This shows that (4.17) implies the estimate (4.16) when γ is large enough.

Next we simplify further the equation. Using (4.4) and replacing the coefficients A_j^ε and E^ε by $\widetilde{A}_j^\sharp$ and $\widetilde{E}^\sharp$, one obtains operators $A^\sharp$ and $M^\sharp$ such that

$$A^\varepsilon = A^\sharp + \varepsilon A^\varepsilon_{d,1}, \quad M^\varepsilon = M^\sharp + \varepsilon \sum_{j=1}^{d-1} A^\varepsilon_{j,1}\varepsilon\partial_j + \varepsilon E_1^\varepsilon. \tag{4.18}$$

Using (4.6), we see that

$$\|A^\varepsilon v - A^\sharp v\|_m \lesssim \varepsilon\|v\|_m, \quad \|M^\varepsilon u - M^\sharp u\|_0 \lesssim \varepsilon\big(\|u\|_m + \|\varepsilon\partial_y u\|_m\big).$$

Therefore, if (U, f') satisfy (4.9), one has

$$\begin{cases} \partial_x U - \dfrac{1}{\varepsilon} G^\sharp U = F \\ \Gamma U_{|x=0} = 0 \end{cases}, \quad G^\sharp := \begin{pmatrix} 0 & \mathrm{Id} \\ M^\sharp & A^\sharp \end{pmatrix}, \quad F = \begin{pmatrix} 0 \\ f, \end{pmatrix} \tag{4.19}$$

with

$$\|f\|_m \lesssim \|f'\|_m + \|u\|_m + \|v\|_m + \|\varepsilon\partial_y u\|_m \lesssim \|f'\|_m + \|u\|_{m,(\Lambda)} + \|v\|_m.$$

Therefore, Theorem 4.1 follows from the following estimate.

THEOREM 4.2. *There are ω, γ_0 and $\varepsilon_0 > 0$, such that the estimate (4.17) is satisfied for all U and f supported in ω and satisfying (4.19).*

The strategy is as follows. We replace the equation (4.19) by a paradifferential equation, at the price of adding a new source term with $\mathcal{H}^m$ norm controlled by $\|u\|_{m,(\Lambda)} + \|v\|_m$. Then we microlocalize the estimate. Next, we prove the estimate with $m = 0$, using symmetrizers which are paradifferential operators whose symbols are given by the constant coefficient analysis of section 2. Finally we prove the conormal estimates, using the special structure of commutators

4.2. Paralinearisation. As in section 2, we introduce for $z \in [0, \infty[$, p in a neighborhood of $\underline{p} = (b(\underline{t}, 0), u_0(\underline{t}, 0), 0)$ and $\zeta = (\tau, \eta, \gamma) \in \mathbb{R}^{d+1}$:

$$\begin{cases} \mathcal{A}(z, p, \zeta) = (B^\sharp_{d,d})^{-1}\Big(A_1^\sharp - \sum_{j=1}^{d-1} i\eta_j(B^\sharp_{j,d} + B^\sharp_{d,j})\Big), \\ \mathcal{M}(z, p, \zeta) = (B^\sharp_{d,d})^{-1}\Big((i\tau + \gamma) + \sum_{j=1}^{d-1} i\eta_j A_j^\sharp + \sum_{j,k=1}^{d-1} \eta_j\eta_k B^\sharp_{j,k} + E^\sharp\Big). \end{cases}$$

Substituting p^ε in place of p, we define

$$a^\varepsilon(t, y, x, \zeta) = \kappa_1(t, y, x)\mathcal{A}(\frac{x}{\varepsilon}, p^\varepsilon(t, y, x), \zeta),$$
$$m^\varepsilon(t, y, x, \zeta) = \kappa_1(t, y, x)\mathcal{M}(\frac{x}{\varepsilon}, p^\varepsilon(t, y, x), \zeta),$$

where $\kappa_1 \in C_0^\infty(\widetilde{\omega})$ and $\kappa = 1$ on a smaller neighborhood $\widetilde{\omega}_1$.

LEMMA 4.3. *The families $\{a^\varepsilon\}$ and $\{m^\varepsilon\}$ are uniformly bounded in $\mathrm{P}\Gamma^1_{1,m}$ and $\mathrm{P}\Gamma^2_{1,m}$ respectively. Moreover,*

$$\|\kappa_1 M^\sharp u - P^{\varepsilon,\gamma}_{m^\varepsilon} u\|_m \lesssim \varepsilon \|u\|_{m,(\Lambda)} ,$$
$$\|\kappa_1 A^\sharp v - P^{\varepsilon,\gamma}_{a^\varepsilon} v\|_m \lesssim \varepsilon \|v\|_m .$$

PROOF. a^ε and m^ε are symbols of differential operators. a^ε is of degree one in η (independent of τ and γ) while m is first order in $\tau - i\gamma$ and second order in η. By (4.5), the coefficients are bounded in $\mathcal{W}^{m,1}$, implying the first statement of the lemma.

The paradifferential operators $P^{\varepsilon,\gamma}$ are those introduced in section 3.2. Recall that $P^{\varepsilon,\gamma}_{ia\eta_j} = P^{\varepsilon,\gamma}_{a}\varepsilon\partial_{y_j}$, $P^{\varepsilon,\gamma}_{ia\tau} = P^{\varepsilon,\gamma}_{a}\varepsilon\partial_t$ and $P^{\varepsilon,\gamma}_{\gamma a} = \varepsilon\gamma P^{\varepsilon,\gamma}_{a}$. The estimates immediately follow from (4.5) and the para-linearization Proposition 3.17. □

Suppose that (U, f) are supported in $\widetilde{\omega}_1$ and satisfy (4.19). Then $G^\sharp U = \kappa_1 G^\sharp U$ and the lemma implies that

$$(4.20)\qquad \begin{cases} \partial_x U - \dfrac{1}{\varepsilon} P^{\varepsilon,\gamma}_{g^\varepsilon} U = F' \\ \Gamma U_{|x=0} = 0 \end{cases} , \qquad g^\varepsilon = \begin{pmatrix} 0 & \mathrm{Id} \\ m^\varepsilon & a^\varepsilon \end{pmatrix}, \quad F' = \begin{pmatrix} 0 \\ f' \end{pmatrix},$$

where f' satisfies

$$(4.21)\qquad \|f'\|_m \lesssim \|f\|_m + \|u\|_{m,(\Lambda)} + \|v\|_m .$$

Suppose that we have a finite collection of symbols independent of (t, y, x), $\chi_l \in \mathrm{P}\Gamma^0$, such that

$$(4.22)\qquad \sum_l \chi_l(\zeta) = 1 \quad \text{on} \quad \mathbb{R}^{d+1}_+ = \{\zeta \in \mathbb{R}^{d+1} \mid \gamma \geq 0\ \}.$$

In addition, suppose that $\kappa \in C^\infty_0(\widetilde{\omega}_1)$ and $\kappa = 1$ on the smaller neighborhood ω. Then, if U is supported in ω, $U = \kappa U$ and

$$U = (\kappa - P^{\varepsilon,\gamma}_\kappa) U + \sum P^{\varepsilon,\gamma}_{\kappa\chi_l} U .$$

Proposition 3.17 and Corollary 3.18 imply that

$$\|(\kappa - P^{\varepsilon,\gamma}_\kappa) u\|_{m,(\Lambda^2)} \lesssim \varepsilon \|u\|_{m,(\Lambda)} , \quad \|(\kappa - P^{\varepsilon,\gamma}_\kappa) v\|_{m,(\Lambda)} \lesssim \varepsilon \|v\|_m .$$

Since $\varphi^2 \lesssim \Lambda^2/\varepsilon$, the definitions (4.13) and (4.15) of weighted norms implies that

$$(4.23)\qquad \|u\|_{m,(\varphi^2)} \leq \frac{1}{\varepsilon} \|u\|_{m,(\Lambda^2)} , \quad \|v\|_{m,(\varphi)} \leq \frac{1}{\sqrt{\varepsilon}} \|v\|_{m,(\Lambda)} .$$

Therefore,

$$\|(\kappa - P^{\varepsilon,\gamma}_\kappa) u\|_{m,(\varphi^2)} \lesssim \|u\|_{m,(\Lambda)} , \quad \frac{1}{\sqrt{\varepsilon}} \|(\kappa - P^{\varepsilon,\gamma}_\kappa) v\|_{m,(\varphi)} \lesssim \|v\|_m .$$

One has similar estimates for the traces. Therefore, the estimate (4.17) and therefore Theorems 4.2 and 4.1 follow from the following microlocal estimates.

PROPOSITION 4.4. *There are a neighborhood ω of $(\underline{t}, 0)$ in $\mathbb{R}^{d+1}_+$, $\kappa \in C^\infty_0(\widetilde{\omega}_1)$ equal to one on ω, a finite partition of unity (4.22) and constants C and $\gamma_0 \geq 1$,* *such that for all $\gamma \geq \gamma_0$, $\varepsilon \in]0,1]$, all l and all solution of (4.19) supported in ω, one has*

$$\|P^{\varepsilon,\gamma}_{\kappa\chi_l} u\|_{m,(\varphi^2)} + \frac{1}{\sqrt{\varepsilon}}\|P^{\varepsilon,\gamma}_{\kappa\chi_l} v\|_{m,(\varphi)} + |P^{\varepsilon,\gamma}_{\kappa\chi_l} v(0)|_{m,(\varphi/\sqrt{\Lambda})} \leq C\Big(\|f\|_m + \|u\|_{m,(\Lambda)} + \|v\|_m + |v(0)|_m\Big). \tag{4.24}$$

As in section 2, there are three different analysis according to the size of $|\varepsilon\zeta|$. We consider successively, the high, medium and low frequencies.

4.3. The high frequency regime. We consider here the case where χ is supported in a domain where Λ is large enough:

$$\chi(\zeta) = 0 \quad \text{when } |\zeta| \leq R+1 \quad \text{and} \quad \chi(\zeta) = 1 \quad \text{when } |\zeta| \geq 2R, \tag{4.25}$$

and we assume that, at least, $R \geq 2$.

As in section 2, in the high frequency regime, it is convenient to reduce the symbol g^ε to first order, so we introduce:

$$g^\varepsilon_2 = \begin{pmatrix} 0 & \Lambda \mathrm{Id} \\ m^\varepsilon \Lambda^{-1} & a^\varepsilon \end{pmatrix}. \tag{4.26}$$

To prove the L^2 estimates, we use the symmetrizers of section 2.

LEMMA 4.5. *There are $R \geq 2$, $c > 0$, a neighborhood ω_1 of $(\underline{t}, 0, 0)$ and a bounded family of self adjoint matrix valued symbols $\sigma^\varepsilon(x) \in \mathrm{P}\Gamma^1_{1,0}$ such that $\varepsilon\partial_x\sigma^\varepsilon(x)$ is bounded in $\mathrm{P}\Gamma^1_{0,0}$ and, for all $U \in \mathbb{C}^{2N}$, all $\varepsilon \in]0,1]$, $(t,y,x) \in \omega_1$ and ζ such that $|\zeta| \geq R$, one has:*

i) $\mathrm{Re}\,\big(\sigma^\varepsilon g^\varepsilon_2 U, U\big) \geq c\Lambda^2\,|U|^2$.

ii) $\mathrm{Re}\,\big(\sigma^\varepsilon(0) U, U\big) \geq c\Lambda\,|v|^2$ *when* $u = 0$.

PROOF. With the notation $\check{\mathcal{G}}_1$ introduced just before Lemma 2.14, there holds

$$g^\varepsilon_2(t,y,x,\zeta) = \Lambda\kappa_1(t,y,x)\,\hat{\mathcal{G}}_1(z, p^\varepsilon, \hat{\zeta}, 1/\Lambda)$$

with $z = x/\varepsilon$, $p^\varepsilon = p^\varepsilon(t,y,x)$, $\Lambda = \Lambda(\zeta)$ and, using the parabolic dilation as in (2.55), $\hat{\zeta} = \Lambda^{-1}\cdot\zeta = (\tau/\Lambda^2, \eta/\Lambda, \gamma/\Lambda^2)$.

Lemma 2.14 provides us with a symmetrizer $\hat{\mathcal{S}}\big(z, p, \hat{\zeta}, \Lambda^{-1}\big)$ for $\hat{\mathcal{G}}_1$, defined for p in some neighborhood of $\underline{p}$ and $\Lambda(\zeta)$ large enough, thus for $|\zeta| \geq R/2$ if R is large. According to Example 3.11 3), the symbols

$$\sigma^\varepsilon(x, \tilde{y}, \zeta) = \kappa_2(t,y,x)\chi_2(\zeta)\Lambda(\zeta)\,\hat{\mathcal{S}}\big(\frac{x}{\varepsilon}, p^\varepsilon(t,y,x), \hat{\zeta}, \frac{1}{\Lambda(\zeta)}\big)$$

are well defined bounded in $\mathrm{P}\Gamma^1_{1,0}$ if $\chi_2 \in C^\infty$ vanishes for $|\zeta| \leq R/2$ and is equal to one on $|\zeta| \geq R$ and $\kappa_2 \in C^\infty$ is supported in a small neighborhood of $(\underline{t}, 0)$ and equal to one on some smaller neighborhood $\omega_1 \subset \widetilde{\omega}_1$. Moreover, the properties of $\hat{\mathcal{S}}$ imply that σ^ε satisfies the desired conditions for ζ large enough. □

The neighborhood ω_1 and R being given by this lemma, we choose χ satisfying (4.25) and $\kappa \in C^\infty_0(\omega_1)$ equal to one on a smaller open set ω. We also choose $\kappa' \in C^\infty_0(\omega_1)$ and χ' supported in $\{|\zeta| \geq R\}$ such that

$$\kappa'\kappa = \kappa, \qquad \chi'\chi = \chi. \tag{4.27}$$

Introduce

$$U_1 = P^{\varepsilon,\gamma}_{\kappa\chi} U = {}^t(u_1, v_1). \tag{4.28}$$

Since $\kappa\chi$ is scalar, the matrices m^ε and a^ε commute with $\kappa\chi\mathrm{Id}$. Thus, by Proposition 3.13, the commutators satisfy

$$\|[P^{\varepsilon,\gamma}_{m^\varepsilon}, P^{\varepsilon,\gamma}_{\kappa\chi}]u\|_m \lesssim \varepsilon\|u\|_{m,(\Lambda)}\,, \quad \|[P^{\varepsilon,\gamma}_{a^\varepsilon}, P^{\varepsilon,\gamma}_{\kappa\chi}]v\|_m \lesssim \varepsilon\|v\|_m\,.$$

We have used that the norms $\|\cdot\|_{m,(\Lambda^s)}$ are the norms denoted by $\|\cdot\|_{m,s,\varepsilon,\gamma}$ in section 3. Hence,

$$\begin{cases} (\partial_x - \frac{1}{\varepsilon}P^{\varepsilon,\gamma}_{g^\varepsilon})U_1 = F_1 = \begin{pmatrix} f_1 \\ g_1 \end{pmatrix}, \\ \Gamma U_{1|x=0} = u_{1|x=0} = 0\,, \end{cases} \tag{4.29}$$

with $f_1 = P^{\varepsilon,\gamma}_{\partial_x\kappa\chi}u$ and $g_1 = f' + P^{\varepsilon,\gamma}_{\partial_x\kappa\chi}v + \frac{1}{\varepsilon}[P^{\varepsilon,\gamma}_{\kappa\chi}, P^{\varepsilon,\gamma}_{m^\varepsilon}]u + \frac{1}{\varepsilon}[P^{\varepsilon,\gamma}_{\kappa\chi}, P^{\varepsilon,\gamma}_{a^\varepsilon}]v$. Thus

$$\|f_1\|_{m,(\Lambda)} + \|g_1\|_m \lesssim \|f'\|_m + \|u\|_{m,(\Lambda)} + \|v\|_m\,. \tag{4.30}$$

The boundary condition in (4.29) follows from the trace relations (3.28).

Next, introduce $U_2 = {}^t(u_2, v_2)$ with

$$u_2 = P^{\varepsilon,\gamma}_{\Lambda}u_1\,, \quad v_2 = v_1\,. \tag{4.31}$$

Then, (4.29) implies that

$$\partial_x U_2 - \frac{1}{\varepsilon}P^{\varepsilon,\gamma}_{g_2^\varepsilon}U_2 = F_2 \qquad u_{2|x=0} = 0\,. \tag{4.32}$$

with $F_2 = {}^t(f_2, g_2)$, $f_2 = P^{\varepsilon,\gamma}_{\Lambda}f_1$ and $g_2 = g_1$. We have used that $u_1 = P^{\varepsilon,\gamma}_{\Lambda^{-1}}u_2$ and $P^{\varepsilon,\gamma}_{m}P^{\varepsilon,\gamma}_{\Lambda^{-1}} = P^{\varepsilon,\gamma}_{m\Lambda^{-1}}$. Because $\varphi \le \Lambda/\sqrt{\varepsilon}$ (indeed $\varphi = \Lambda/\sqrt{\varepsilon}$ on the support of χ), the estimate (4.24) for $U_1 = P^{\varepsilon,\gamma}_{\kappa\chi}U$ is implied by

$$\begin{aligned} \frac{1}{\varepsilon}\|U_2\|_{m,(\Lambda)} + \frac{1}{\sqrt{\varepsilon}}|v_2(0)|_{m,(\sqrt{\Lambda})} \lesssim \\ \|F_2\|_m + \|u\|_{m,(\Lambda)} + \|v\|_m + |v(0)|_m\,, \end{aligned} \tag{4.33}$$

since $\|F_2\|_m \lesssim \|f'\|_m + \|u\|_{m,(\Lambda)} + \|v\|_m$ by (4.30).

PROPOSITION 4.6. *If R is large enough, for all $\gamma \ge 1$, all $\varepsilon \in]0,1]$ and all $(U,f) \in C_0^\infty(\omega)$ solution of (4.19), one has*

$$\begin{aligned} \frac{1}{\varepsilon}\|P^{\varepsilon,\gamma}_{\kappa'\chi'}U_2\|_{0,(\Lambda)} + \frac{1}{\sqrt{\varepsilon}}|P^{\varepsilon,\gamma}_{\kappa'\chi'}v_2(0)|_{0,(\sqrt{\Lambda})} \lesssim \\ \|F_2\|_0 + \|U_2\|_0 + |v_2(0)|_0\,. \end{aligned} \tag{4.34}$$

PROOF. Introduce $U_2' = P^{\varepsilon,\gamma}_{\kappa'\chi'}U_2$. Then, using the symbolic calculus,

$$\partial_x U_2' - \frac{1}{\varepsilon}P^{\varepsilon,\gamma}_{g_2^\varepsilon}U_2' = F_2'\,, \qquad u'_{2|x=0} = 0\,,$$

with

$$\|F_2'\|_0 \lesssim \|F_2\|_0 + \|U_2\|_0\,.$$

Consider the symmetrizers $S = \mathrm{Re}\,P^{\varepsilon,\gamma}_{\sigma^\varepsilon} = \frac{1}{2}(P^{\varepsilon,\gamma}_{\sigma^\varepsilon} + (P^{\varepsilon,\gamma}_{\sigma^\varepsilon})^*)$. By Proposition 3.14, since the symbols $\sigma^\varepsilon \in P\Gamma^1_{1,0}$ are self adjoint, $S - P^{\varepsilon,\gamma}_{\sigma^\varepsilon}$ is of order 0 in the sense of section 3.2. The following identity holds:

$$\begin{aligned} (\!(S(0)U_2'(0), U_2'(0))\!) + \frac{2}{\varepsilon}\mathrm{Re}\,(\!(SG_2U_2', U_2')\!) = \\ - (\!([\partial_x, S]U_2', U_2')\!) - 2\mathrm{Re}\,(\!(SF_2', U_2')\!), \end{aligned}$$

where $G_2 = P^{\varepsilon,\gamma}_{g_2^\varepsilon}$. The symbolic calculus in Proposition 3.13 implies that $SG_2 = P^{\varepsilon,\gamma}_{\sigma^\varepsilon g_2^\varepsilon} + \varepsilon R$ with R of order one. Thus

$$\big(\!\big(SG_2U_2', U_2'\big)\!\big) = \big(\!\big(P^{\varepsilon,\gamma}_{\sigma^\varepsilon g_2^\varepsilon} U_2', U_2'\big)\!\big) + O\Big(\varepsilon \, \|U_2'\|_{0,(\Lambda)} \|U_2'\|_0\Big).$$

Moreover, $\sigma^\varepsilon g_2^\varepsilon \in P\Gamma^2_{1,0}$ and its real part is elliptic on the support of $\kappa\chi$. Thus, with Remark 3.16, we are in position to apply Garding's inequality of Proposition 3.15, and therefore

$$\mathrm{Re}\, \big(\!\big(SG_2U_2', U_2'\big)\!\big) \geq c\|U_2'\|^2_{0,(\Lambda)} - C\,\varepsilon^2 \|U_2\|_0^2 \,.$$

Similarly, since $u_2'(0) = 0$,

$$\big(\!\big(S(0)U_2'(0), U_2'(0)\big)\!\big) \geq c|v_2'(0)|^2_{0,(\sqrt{\Lambda})} - C\varepsilon^2 |v_2(0)|^2_{0,(1/\sqrt{\Lambda})} \,.$$

Using the symbolic calculus, we also have

$$\big|\big(\!\big(SF_2', U_2'\big)\!\big)\big| \leq \|F_2'\|_0 \, \|S^*U_2'\|_0 \leq \frac{\beta}{\varepsilon} \, \|U_2'\|^2_{0,(\Lambda)} + \frac{C\varepsilon}{\beta} \, \|F_2\|_0^2$$

for all $\beta > 0$. We have used that S and thus S^* are of order one.

Since $\varepsilon\partial_x\sigma^\varepsilon$ is bounded in $P\Gamma^1_{0,0}$, $[\varepsilon\partial_x, P^{\varepsilon,\gamma}_{\sigma^\varepsilon}]$ is of order 1, as well as its adjoint. Thus $\varepsilon[\partial_x, S]$ is also of order 1 and

$$\big|\big(\!\big([\partial_x, S]U_2', U_2')\big)\!\big)\big| \leq \frac{C_1}{\varepsilon} \, \|U_2'\|^2_{0,(\sqrt{\Lambda})} \,.$$

This implies that

$$\begin{aligned}\frac{1}{\varepsilon}\|U_2'\|^2_{0,(\Lambda)} + |v_2'(0)|^2_{0,(\sqrt{\Lambda})} \leq & C_1 \, \frac{1}{\varepsilon}\|U_2'\|^2_{0,(\sqrt{\Lambda})} + \frac{\beta}{\varepsilon} \, \|U_2'\|^2_{0,(\Lambda)} \\ & + C_2\big(\frac{\varepsilon}{\beta}\|F_2'\|_0^2 + \varepsilon\|U_2\|_0^2 + \varepsilon^2|v(0)|_0^2\big) \,.\end{aligned} \tag{4.35}$$

We first choose $\beta \leq 1/4$ so that the second term in the right hand side can be absorbed from the right to the right.

Next, we note that the constant C_1 depends only on norms of the symbols σ^ε. In particular, it is independent χ. If R is sufficiently large in (4.25), the symbol $\kappa'\chi'$ is supported in $\langle\zeta\rangle \geq \sqrt{R}/2$. Thus, by *iv)* in Proposition 3.12, the spectrum of U_2' is also contained in a domain $\langle\varepsilon\zeta\rangle \geq \delta\sqrt{R}$ with $\delta > 0$. If R is large enough, one has $C_1 \leq \frac{1}{4}\sqrt{\Lambda}(\varepsilon\zeta)$ on the spectrum of U_2' and thus

$$4C_1\|U_2\|_{0,(\sqrt{\Lambda})} \leq \|U_2\|_{0,(\Lambda)} \,.$$

Therefore, the first term in the right hand side of (4.35) can be also absorbed by the left hand side, implying (4.34). □

PROPOSITION 4.7. *If R is large enough, there are C and $\gamma_0 \geq 1$ such that the estimate* (4.24) *is satisfied by $P^{\varepsilon,\gamma}_{\kappa\chi}U$, for all $\gamma \geq \gamma_0$, all $\varepsilon \in]0,1]$ and all $(U, f) \in C_0^\infty(\omega)$ satisfying* (4.19).

PROOF. Differentiate (4.32) with respect to Z^α, for $|\alpha| \leq m$. The commutation relation (3.20) implies that

$$[Z^\alpha, P^{\varepsilon,\gamma}_{g_2^\varepsilon}] = \sum_{\beta<\alpha} \tbinom{\alpha}{\beta} P^{\varepsilon,\gamma}_{Z^{\alpha-\beta}g_1^\varepsilon} Z^\beta \,.$$

Since g_1^ε is bounded in $\mathrm{P}\Gamma^1_{1,m}$, the following estimates hold for $w \in P\mathcal{H}^{m,1}$:

$$\|[Z^\alpha, P^{\varepsilon,\gamma}_{g_2^\varepsilon}]w\|_0 \lesssim \sum_{\beta<\alpha} \|Z^\beta w\|_{0,(\Lambda)}\,.$$

Thus,

$$\gamma^{m-|\alpha|}\|[Z^\alpha, P^{\varepsilon,\gamma}_{g_2^\varepsilon}]U_2\|_0 \lesssim \gamma^{-1}\|U_2\|_{m,(\Lambda)}\,.$$

Therefore,

$$\begin{cases} \partial_x Z^\alpha U_2 - \dfrac{1}{\varepsilon} P^{\varepsilon,\gamma}_{g_2^\varepsilon} Z^\alpha U_1 = Z^\alpha F_2 + \dfrac{1}{\varepsilon} F_\alpha\,, \\ Z^\alpha u_{2|x=0} = 0\,, \end{cases}$$

where

$$\gamma^{m-|\alpha|}\|F_\alpha\|_0 \lesssim \gamma^{-1}\|U_2\|_{m,(\Lambda)}\,.$$

Proposition 4.6 implies that

$$\begin{aligned} \frac{1}{\varepsilon}\|P^{\varepsilon,\gamma}_{\kappa'\chi'} Z^\alpha U_2\|_{0,(\Lambda)} + \frac{1}{\sqrt{\varepsilon}}|P^{\varepsilon,\gamma}_{\kappa'\chi'} Z^\alpha v_1(0)|_{0,(\sqrt{\Lambda})} \ \lesssim \\ \|Z^\alpha F_2\|_0 + \frac{1}{\varepsilon}\|F_\alpha\|_0 + \|Z^\alpha U_2\|_0 + |Z^\alpha v_0(0)|_0\,. \end{aligned} \tag{4.36}$$

We use the following Lemma.

LEMMA 4.8. *Suppose that* $a \in \mathrm{P}\Gamma^0_{1,m}$ *and* $b \in \Gamma^0_{1,m}$ *satisfy* $ab = b$*. Then, with* $h_2 = P^{\varepsilon,\gamma}_b h$*, one has*

$$\|h_2\|_{m,(\Lambda)} \lesssim \sum_{|\alpha|\le m} \gamma^{m-|\alpha|}\|P^{\varepsilon,\gamma}_a Z^\alpha h_2\|_{0,(\Lambda)} + \varepsilon\|h\|_m\,.$$

Taking this lemma for granted, since $U_2 = P^{\varepsilon,\gamma}_{\kappa'\chi'} U'$ with $u'_1 = P^{\varepsilon,\gamma}_\Lambda u$ and $v' = v$, with (4.27) and (4.36) and the estimate of F_α, we obtain

$$\begin{aligned} \frac{1}{\varepsilon}\|U_2\|_{m,(\Lambda)} + \frac{1}{\sqrt{\varepsilon}}|v_1(0)|_{m,(\sqrt{\Lambda})} \ \lesssim \\ \frac{1}{\varepsilon\gamma}\|U_2\|_{m,(\Lambda)} + \|F_2\|_m + \|U'\|_m + |v(0)|_m\,. \end{aligned}$$

For γ large enough, this implies (4.33) and thus (4.24) for $P^{\varepsilon,\gamma}_{\kappa_1\chi} U$. □

PROOF OF LEMMA 4.8. The symbolic calculus implies that $P^{\varepsilon,\gamma}_a h_2 = P^{\varepsilon,\gamma}_{ab} h + \varepsilon R$ where R is of order -1. Because $ab = b$, this shows that $P^{\varepsilon,\gamma}_a h_2 = h_2 + \varepsilon R h$. Thus

$$\|h_2\|_{m,(\Lambda)} \lesssim \sum_{|\alpha|\le m} \gamma^{m-|\alpha|}\|Z^\alpha P^{\varepsilon,\gamma}_a h_2\|_{0,(\Lambda)} + \varepsilon\|h\|_m.$$

Next we commute Z^α and $P^{\varepsilon,\gamma}_a$: $[Z^\alpha P^{\varepsilon,\gamma}_a]h_2$ is a sum of terms

$$k = P^{\varepsilon,\gamma}_{Z^\beta a} P^{\varepsilon,\gamma}_{Z^{\beta'} b} Z^{\alpha-\beta-\beta'} h$$

with $\beta > 0$ and $\beta + \beta' \le \alpha$. Because $a = 1$ on the support of b, for $\beta > 0$, $(Z^\beta a)(Z^{\beta'} b) = 0$, and the symbolic calculus implies that

$$\gamma^{m-|\alpha|}\|k\|_{0,(\Lambda)} \lesssim \varepsilon\gamma^{m-|\alpha|}\|Z^{\alpha-\beta-\beta'} h\|_0 \le \varepsilon\|h\|_m\,.$$

The lemma follows. □

4.4. Bounded frequencies. From now on, we consider bounded frequencies ζ. Fix $\underline{\zeta} = (\underline{\tau}, \underline{\eta}, \underline{\gamma}) \in \mathbb{R}^{d+1}$ with $\underline{\gamma} \geq 0$ and $\underline{\zeta} \neq 0$. Lemma 2.6 provides us with invertible matrices $\mathcal{W}(z, p, \zeta)$ defined for (p, ζ) close to $(\underline{p}, \underline{\zeta})$ and such that

$$\partial_z \mathcal{W} + \mathcal{W}\mathcal{G} = \mathcal{G}_2 \mathcal{W}\,, \tag{4.37}$$

where $\mathcal{G}_2(p, \zeta)$ is independent of z. Lemma 2.12 provides us with a symmetrizer $\mathcal{S}(p, \zeta)$ for $\mathcal{G}_2$ in a neighborhood of $(\underline{p}, \underline{\zeta})$. Introduce

$$\begin{aligned} g_2^\varepsilon(t, y, x, \zeta) &= \kappa_2(t, y, x)\, \chi_2(\zeta)\mathcal{G}_2(p^\varepsilon(x, \tilde{y}), \zeta)\,, \\ \sigma^\varepsilon(t, y, x, \zeta) &= \kappa_3(t, y, x)\chi_3(\zeta)\mathcal{S}(p^\varepsilon(t, y, x), \zeta)\,, \\ w^\varepsilon_{-1}(t, y, x, \zeta) &= \kappa_2(t, y, x)\, \chi_2(\zeta)\mathcal{W}^{-1}(\frac{x}{\varepsilon}, p^\varepsilon(t, y, x), \zeta)\,, \\ \widetilde{\Gamma}^\varepsilon(t, y, \zeta) &= \Gamma w^\varepsilon_{-1}(t, y, 0, \zeta)\,, \end{aligned}$$

with C^∞ cut-offs κ_2, κ_3 and χ_2, χ_3, supported on small neighborhoods of $(\underline{t}, 0)$ and $\underline{\zeta}$ respectively. We further assume that $\kappa_2\kappa_3 = \kappa_3$, $\chi_2\chi_3 = \chi_3$, and that p^ε remains in the domain of definition of $\mathcal{G}_2, \mathcal{S}$ and $\mathcal{W}$ for (t, y, x) in the support of κ_2. We also assume that $\kappa_3 = 1$ and $\chi_3 = 1$ on smaller neighborhoods ω_1, and C_1. Thanks to (4.5), these symbols are bounded in $\mathrm{P}\Gamma^0_{1,m}$ for $\varepsilon \in]0, 1]$. Note that the order, has no real meaning here, since the symbols are compactly supported in ζ. Moreover, since $\mathcal{S}$ does not depend on z, the family $\partial_x\sigma^\varepsilon$ is also bounded in $\mathrm{P}\Gamma^0_{0,m}$. Lemma 2.12 implies:

LEMMA 4.9. *With notations as above, there are constants C and $c > 0$ such that the self adjoint matrices σ^ε for $\varepsilon \in]0, 1]$ form a bounded family in $\mathrm{P}\Gamma^0_{1,0}$ with $\partial_x\sigma^\varepsilon$ bounded in $\mathrm{P}\Gamma^0_{0,0}$ and for all $U \in \mathbb{C}^{2N}$, $\varepsilon \in]0, 1]$, $(t, y, x) \in \omega_1$ and $\zeta \in \mathrm{C}_1$:*
i) $\mathrm{Re}\,(\sigma^\varepsilon g_2^\varepsilon U, U) \geq c|U|^2$.
ii) $\mathrm{Re}\,(\sigma^\varepsilon(0)U, U) + C|\widetilde{\Gamma}^\varepsilon U|^2 \geq c|U|^2$.

With $\kappa \in C_0^\infty(\omega_1)$ equal to one on ω and $\chi \in C_0^\infty(\mathrm{C}_1)$ equal to one on C, introduce

$$w^\varepsilon(t, y, x, \zeta) = \kappa(t, y, x)\, \chi(\zeta)\mathcal{W}(\frac{x}{\varepsilon}, p^\varepsilon(t, y, x), \zeta)\,. \tag{4.38}$$

The intertwining relation (4.37) is transformed to

$$\varepsilon\partial_x w^\varepsilon + w^\varepsilon g^\varepsilon = g_2^\varepsilon w^\varepsilon + \varepsilon r^\varepsilon \tag{4.39}$$

with $r^\varepsilon = \partial_x\kappa\chi\mathcal{W} + \kappa\chi\partial_x p\nabla_p\mathcal{W}$ bounded in $\mathrm{P}\Gamma^0_{0,m}$.

Consider $U_2 = P^{\varepsilon,\gamma}_{w^\varepsilon}U$. Then, with (4.20) and (4.39) and using the symbolic calculus, we obtain

$$\begin{aligned} \partial_x U_2 &= P^{\varepsilon,\gamma}_{w^\varepsilon}\partial_x U + P^{\varepsilon,\gamma}_{\partial_x w^\varepsilon}U \\ &= \frac{1}{\varepsilon}P^{\varepsilon,\gamma}_{g_2^\varepsilon w^\varepsilon}U + \frac{1}{\varepsilon}(P^{\varepsilon,\gamma}_{w^\varepsilon}P^{\varepsilon,\gamma}_{g^\varepsilon} - P^{\varepsilon,\gamma}_{w^\varepsilon g^\varepsilon})U + P^{\varepsilon,\gamma}_{r^\varepsilon}U + P^{\varepsilon,\gamma}_{w^\varepsilon}F'\,. \end{aligned}$$

But $\varepsilon^{-1}(P^{\varepsilon,\gamma}_{g_2^\varepsilon w^\varepsilon} - P^{\varepsilon,\gamma}_{g_2^\varepsilon}P^{\varepsilon,\gamma}_{w^\varepsilon})$ and $\varepsilon^{-1}(P^{\varepsilon,\gamma}_{w^\varepsilon g^\varepsilon} - P^{\varepsilon,\gamma}_{w^\varepsilon}P^{\varepsilon,\gamma}_{g^\varepsilon})$ are of order zero, since w^ε is of any negative order, being compactly supported in ζ. Here, the important fact is not about the order of the operators, but that the remainders in the symbolic calculus are smaller by a factor ε. Thus

$$\partial_x U_2 = \frac{1}{\varepsilon}P^{\varepsilon,\gamma}_{g_2^\varepsilon}U_2 + F_2\,, \tag{4.40}$$

where F_2 satisfies

$$\|F_2\|_m \lesssim \|f'\|_m + \|U\|_m \,. \tag{4.41}$$

Next, the symbols satisfy $w^\varepsilon_{-1} w^\varepsilon = \kappa\chi \mathrm{Id}$. Thus,

$$P^{\varepsilon,\gamma}_{\kappa\chi} U = P^{\varepsilon,\gamma}_{w^\varepsilon_{-1}} U_2 + \varepsilon R U \,, \tag{4.42}$$

where R is of order -1. Thus

$$\|P^{\varepsilon,\gamma}_{\kappa\chi} U\|_m \lesssim \|U_2\|_m + \varepsilon \|U\|_m \,. \tag{4.43}$$

The boundary condition $u(0) = 0$ and (4.42) imply that

$$\big|P^{\varepsilon,\gamma}_{\widetilde{\Gamma}^\varepsilon} U_2(0)\big|_m \lesssim \varepsilon \big|R(0) v(0)\big|_m \lesssim \varepsilon \big|v(0)\big|_m \,. \tag{4.44}$$

Similarly, (4.42) implies that

$$\big|P^{\varepsilon,\gamma}_{\kappa\chi} v(0)\big|_m \lesssim \big|U_2(0)\big|_m + \varepsilon \big|v(0)\big|_m \,. \tag{4.45}$$

Moreover, since w^ε and χ are compactly supported, the spectrum of $P^{\varepsilon,\gamma}_{\kappa\chi}$ is contained in a domain where $\varepsilon\zeta$ is uniformly bounded. Hence $\varphi \lesssim 1/\sqrt{\varepsilon}$ on this domain and

$$\|P^{\varepsilon,\gamma}_{\kappa\chi} u\|_{m,(\varphi^2)} + \frac{1}{\sqrt{\varepsilon}} \|P^{\varepsilon,\gamma}_{\kappa\chi} v\|_{m,(\varphi)} \lesssim \frac{1}{\varepsilon} \|P^{\varepsilon,\gamma}_{\kappa\chi} U\|_m \,.$$

Similarly,

$$|P^{\varepsilon,\gamma}_{\kappa\chi} v(0)|_{m,(\varphi/\sqrt{\Lambda})} \lesssim \frac{1}{\varepsilon} |P^{\varepsilon,\gamma}_{\kappa\chi} v(0)|_m \,.$$

Therefore, with (4.43) (4.45), we conclude that the estimate (4.24) for $P^{\varepsilon,\gamma}_{\kappa\chi} U$ is implied by

$$\frac{1}{\varepsilon} \|U_2\|_m + \frac{1}{\sqrt{\varepsilon}} |U_2(0)|_{m,} \lesssim \|F_2\|_m + \|U\|_m + |v(0)|_m \,. \tag{4.46}$$

PROPOSITION 4.10. *With notations as above, there are C and $\gamma_0 \geq 1$ such that the estimate (4.24) is satisfied by $P^{\varepsilon,\gamma}_{\kappa\chi} U$ for all $\gamma \geq \gamma_0$, all $\varepsilon \in]0,1]$ and all $u \in C^\infty_0(\overline{\omega})$.*

PROOF. We prove (4.46).

a) Choose $\kappa'_1 \in C^\infty_0(\omega_1)$ and $\chi' \in C^\infty_0(\mathcal{C})$ such that $\kappa'\kappa = \kappa$, $\chi'\chi = \chi$ and introduce $U'_2 = P^{\varepsilon,\gamma}_{\kappa'_1\chi'} U_2$. Then, using the symbolic calculus,

$$\|\partial_x U'_2 - \frac{1}{\varepsilon} P^{\varepsilon,\gamma}_{g^\varepsilon_2} U'_2\|_0 \lesssim \|F_2\|_0 + \|U_2\|_0 \,, \qquad |P^{\varepsilon,\gamma}_{\widetilde{\Gamma}^\varepsilon} U'_2(0)|_0 \lesssim |P^{\varepsilon,\gamma}_{\widetilde{\Gamma}^\varepsilon} U_2(0)|_0 \,.$$

Consider the symmetrizers $S = \operatorname{Re} P^{\varepsilon,\gamma}_{\sigma^\varepsilon} = \frac{1}{2}(P^{\varepsilon,\gamma}_{\sigma^\varepsilon} + (P^{\varepsilon,\gamma}_{\sigma^\varepsilon})^*)$. We now repeat the proof of Proposition 4.6 using the symmetrizer $S = \operatorname{Re} P^{\varepsilon,\gamma}_{\sigma^\varepsilon}$. The main difference is that $[\partial_x, S]$ is now a term of order 0, since

$$[\partial_x, S] = \frac{1}{2} P^{\varepsilon,\gamma}_{\partial_x \sigma^\varepsilon} + \frac{1}{2} \big(P^{\varepsilon,\gamma}_{\partial_x \sigma^\varepsilon}\big)^*$$

and $\partial_x \sigma^\varepsilon$ is bounded in $P\Gamma^0_{0,0}$. Therefore, there is now no term in $C_1 \varepsilon^{-1} \|U'_2\|_0$ in the right hand side of (4.35). In addition, *ii)* in Lemma 4.9 implies that

$$\big(\!\big(S(0) U'_2(0), U'_2(0)\big)\!\big) + C |P^{\varepsilon,\gamma}_{\widetilde{\Gamma}^\varepsilon} U_2(0)|_0 \geq c |U'_2(0)|^2_0 - C' \varepsilon^2 |U_2(0)|^2_0 \,.$$

Adding up yields

$$\frac{1}{\varepsilon} \|P^{\varepsilon,\gamma}_{\kappa'_1\chi'} U_2\|_0 + \frac{1}{\sqrt{\varepsilon}} |P^{\varepsilon,\gamma}_{\kappa'_1\chi'} U_2(0)|_0 \lesssim \|F_2\|_0 + \|U_2\|_0 + |U_2(0)|_0 \,.$$

b) The second part of the proof is identical to the proof of Proposition 4.7. We differentiate the equation (4.40) with respect to Z^α. Using part a), one obtains:

$$\frac{1}{\varepsilon}\|P^{\varepsilon,\gamma}_{\kappa_1'\chi'}Z^\alpha U_2\|_0 + \frac{1}{\sqrt{\varepsilon}}|P^{\varepsilon,\gamma}_{\kappa_1'\chi'}Z^\alpha U_2(0)|_0 \lesssim \|F_2\|_m + (1+\frac{1}{\varepsilon\gamma})\|U_2\|_m + |U_2(0)|_m \,.$$

With Lemma 4.8, this implies

$$\frac{1}{\varepsilon}\|U_2\|_m + \frac{1}{\sqrt{\varepsilon}}|U_2(0)|_m \lesssim \|F_2\|_m + \frac{1}{\varepsilon\gamma}\|U_2\|_m + \|U\|_m + |U(0)|_m \,.$$

For γ large, this implies (4.46), finishing the proof of the proposition. □

4.5. The low frequencies. We now turn to the most delicate part, where we consider frequencies supported in a small neighborhood of $\underline{\zeta} = 0$. We start the analysis as in the previous subsection. With $\mathcal{W}$ and $\mathcal{V}$ given by Lemmas 2.6 and 2.9, we define $\mathcal{T} = \mathcal{W}\mathcal{V}$. Then, (4.37) reads

$$\partial_z\mathcal{T} + \mathcal{T}\mathcal{G} = \mathcal{G}_2\mathcal{T}\,, \quad \mathcal{G}_2 = \begin{pmatrix} H & 0 \\ 0 & P \end{pmatrix}.$$

We define

$$\begin{aligned} g_2^\varepsilon(t,y,x,\zeta) &= \kappa_2(t,y,x)\,\chi_2(\zeta)\mathcal{G}_2(p^\varepsilon(x,\tilde{y}),\zeta)\,, \\ w_{-1}^\varepsilon(t,y,x,\zeta) &= \kappa_2(t,y,x)\,\chi_2(\zeta)\mathcal{T}^{-1}(\frac{x}{\varepsilon},p^\varepsilon(t,y,x),\zeta)\,, \\ \widetilde{\Gamma}^\varepsilon(t,y,\zeta) &= \Gamma w_{-1}^\varepsilon(t,y,0,\zeta)\,, \end{aligned}$$

with C^∞ cut-offs κ_2 and χ_2 supported on small neighborhoods of $(\underline{t},0)$ and $\underline{\zeta}$ respectively and equal to one on smaller neighborhoods. Now g_2^ε is block diagonal:

$$g_2^\varepsilon = \begin{pmatrix} h^\varepsilon & 0 \\ 0 & \pi^\varepsilon \end{pmatrix}.$$

Again we use symmetrizers but we now have to combine the $P^{\varepsilon,\gamma}$ calculus for the second block and the T^γ calculus for the first one. We start with some preparation. By definition,

$$h^\varepsilon(t,y,x,\zeta) = \kappa_2(t,y,x)\chi_2(\zeta)H(p^\varepsilon(t,y,x),\zeta)\,.$$

Moreover, H vanishes at the origin. Therefore, using a Taylor expansion of H, we obtain

$$h^\varepsilon = i\tau h_0^\varepsilon + \sum_{j=1}^{d-1} i\eta_j h_j^\varepsilon + \gamma h_d^\varepsilon\,, \tag{4.47}$$

where the h_k^ε are bounded families of symbols in $\mathrm{P}\Gamma^0_{1,0}$, with compact support in ζ. Introduce the symbols

$$\begin{aligned} \mathrm{h}_k^\varepsilon(t,y,x,\check{\zeta}) &:= h_k^\varepsilon(t,y,x,\varepsilon\check{\zeta})\,, \\ \mathrm{h}^\varepsilon(t,y,x,\check{\zeta}) &:= i\tau\mathrm{h}_0^\varepsilon(t,y,x,\check{\zeta}) + \sum_{j=1}^{d-1} i\eta_j\mathrm{h}_j^\varepsilon(t,y,x,\check{\zeta}) + \gamma\mathrm{h}_d^\varepsilon(t,y,x,\check{\zeta})\,. \end{aligned}$$

Then, since the h_k have compact support in ζ, the h_k^ε are bounded families in $\Gamma^0_{1,0}$ and the h^ε are bounded in $\Gamma^1_{1,0}$ (see Example 3.3).

We now proceed to the construction of symmetrizers.

LEMMA 4.11. *There are neighborhoods ω_1 of $(\underline{t}, 0)$ and C_1 of the origin in $\mathbb{R}^{d+1}_+$ and there are C and $c > 0$ such that:*

i) there is a bounded family of self adjoint matrix valued symbols $\sigma_2^\varepsilon \in P\Gamma^0_{1,0}$ such that $\partial_x \sigma_2^\varepsilon$ is bounded in $P\Gamma^0_0$ and, for all $v \in \mathbb{C}^N$, all $\varepsilon \in]0,1]$, $(t,y,x) \in \omega_1$ and $\zeta \in \mathrm{C}_1$, one has $\mathrm{Re}\,(\sigma_2^\varepsilon \pi^\varepsilon v, v) \geq c\|v\|^2$.

ii) there is a bounded family of self adjoint matrix valued symbols $\mathrm{s}_1^\varepsilon \in \Gamma^0_{1,0}$ such that $\partial_x \mathrm{s}_1^\varepsilon$ is bounded in $\Gamma^0_{0,0}$ and

$$\mathrm{Re}\,(\mathrm{s}_1^\varepsilon \mathrm{h}^\varepsilon) = \sum (\mathrm{v}_l^\varepsilon)^* \mathrm{k}_l^\varepsilon \mathrm{v}_l^\varepsilon \tag{4.48}$$

where

a) the $\mathrm{v}^\varepsilon(t,y,x,\check\zeta)$ are bounded families matrix valued symbols in Γ^0_1 such that $\sum (\mathrm{v}_l^\varepsilon)^* \mathrm{v}_l^\varepsilon \geq c\mathrm{Id}$ when $(t,y,x) \in \omega_1$ and $\varepsilon\check\zeta \in \mathrm{C}_1$,

b) the $\mathrm{k}_l^\varepsilon(t,y,x,\check\zeta)$ are bounded families of matrix valued symbols in Γ^1_1 having the following block structure:

$$\mathrm{k}_l^\varepsilon = \begin{bmatrix} \mathrm{b}_1^\varepsilon & \cdots & 0 \\ \vdots & \ddots & \vdots \\ 0 & \cdots & \mathrm{b}_q^\varepsilon \end{bmatrix} \tag{4.49}$$

and either

$$\mathrm{b}_j^\varepsilon(t,y,x,\check\zeta) \geq c|\check\zeta|$$

or $\mathrm{b}_j^\varepsilon = \gamma \mathrm{b}_{j,0} + \varepsilon \mathrm{b}_{j,2}$ and

$$\mathrm{b}_{j,0}^\varepsilon(t,y,x,\check\zeta) \geq c \quad \text{and} \quad \mathrm{b}_{j,2}^\varepsilon(t,y,x,\check\zeta) \geq c|\check\zeta|^2$$

when $(t,y,x) \in \omega_1$ and $\varepsilon\check\zeta \in \mathrm{C}_1$.

iii) the matrices $\mathrm{s}^\varepsilon = \begin{bmatrix} \mathrm{s}_1^\varepsilon & 0 \\ 0 & \mathrm{s}_2^\varepsilon \end{bmatrix}$ with $\mathrm{s}_2^\varepsilon(t,y,x,\check\zeta) = \sigma_2^\varepsilon(x,\tilde y,\varepsilon\check\zeta)$ and $\mathrm{g}^\varepsilon(t,y,\check\zeta) := \widetilde{\Gamma}^\varepsilon(t,y,\varepsilon\check\zeta)$ satisfy for $(t,y) \in \omega_1 \cap \{x = 0\})$ and $\varepsilon\check\zeta \in \mathrm{C}_1$

$$(\!(\mathrm{s}^\varepsilon(0)U, U)\!) + C|\mathrm{g}^\varepsilon U|^2 \geq c|U|^2 . \tag{4.50}$$

PROOF. This is a direct consequence of Lemma 2.13. In this lemma, two symmetrizers $\check S_1(p,\check\zeta,\rho)$ and $S_2(p,\zeta)$ are constructed for the blocks $\check H$ and E respectively. Define

$$\begin{aligned} \mathrm{s}_1^\varepsilon(t,y,x,\check\zeta) &= \kappa_3(t,y,x)\,\chi_3(\varepsilon\check\zeta)\,\check S(p^\varepsilon(t,y,x)\frac{\check\zeta}{|\check\zeta|}, \varepsilon|\check\zeta|) \\ \sigma_3^\varepsilon(t,y,x,\zeta) &= \kappa_2(t,y,x)\,\chi_3(\zeta) S_2(p^\varepsilon(t,y,x),\zeta) \end{aligned} \tag{4.51}$$

with κ_3 and χ_3 appropriate cut off functions near $(\underline{t}, 0)$ and $\underline{\zeta} = 0$, such that $\kappa_2\kappa_3 = \kappa_3$, $\chi_2\chi_3 = \chi_3$ and κ_3 on ω_1 and $\chi_3 = 1$ on C_1. Then, Lemma 2.13 implies the properties *i*), *ii*) and *iii*) above. □

Given ω_1 and C_1, we choose $\kappa \in C_0^\infty(\omega_1)$ and $\chi \in C_0^\infty(\mathrm{C}_1)$ with $\kappa_1 = 1$ and $\chi = 1$ on smaller neighborhoods ω and C. With

$$w^\varepsilon(t,y,x,\zeta) = \kappa(t,y,x)\,\chi(\zeta)\mathcal{T}(\frac{x}{\varepsilon}, p^\varepsilon(t,y,x),\zeta) ,$$

we first give estimates for $U_2 = P^{\varepsilon,\gamma}_{w^\varepsilon} U$. With these notations, the intertwining relation (4.39) still holds, as well as equation (4.40) which now decouples in two parts

$$\partial_x u_2 = \tfrac{1}{\varepsilon} P^{\varepsilon,\gamma}_{h^\varepsilon} u_2 + f_2 \,, \tag{4.52}$$

$$\partial_x v_2 = \tfrac{1}{\varepsilon} P^{\varepsilon,\gamma}_{\pi^\varepsilon} v_2 + g_2 \,, \tag{4.53}$$

where $U_2 = {}^t(u_2, v_2)$ and $F_2 = {}^t(f_2, g_2)$ satisfies (4.41).

We also choose $\kappa' \in C_0^\infty(\omega_1)$ and $\chi' \in C_0^\infty(\mathrm{C}_1)$ such that $\kappa'\kappa = \kappa$ and $\chi'\chi' = \chi$. We introduce $U_2' = P^{\varepsilon,\gamma}_{\kappa_0'\chi'} U_2 = {}^t(u_2', v_2')$. These functions have their spectrum contained in a domain where $\varepsilon\zeta$ is bounded (see Proposition 3.12). In this case $\varphi^2 \approx \gamma + \varepsilon|\tilde\eta|^2$ and for such functions

$$\|u\|_{0,(\varphi^2)} \approx \gamma \|u\|_0 + \varepsilon \|\partial^2_{\tilde y} u\|_0 \,, \qquad \|v\|_{0,(\varphi)} \approx \sqrt{\gamma}\|v\|_0 + \sqrt{\varepsilon}\|\partial_{\tilde y} v\|_0 \,. \tag{4.54}$$

PROPOSITION 4.12. *There is C such that for all $u \in C_0^\infty(\overline{\omega})$*

$$\|u_2'\|_{0,(\varphi^2)} + \frac{1}{\sqrt{\varepsilon}} \|v_2'\|_{0,(\varphi)} + |U_2'(0)|_{0,(\varphi)} \le C\Big(\|f_2\|_0 + \sqrt{\varepsilon}\|g_2\|_{0,(\varphi)} + |P^{\varepsilon,\gamma}_{\widetilde\Gamma^\varepsilon} U_2(0)|_{0,(\varphi)} + \|u_2\|_{0,(\varphi)} + \|v_2\|_0 + |U_2(0)|_0 \Big) \,.$$

PROOF. Applying $P^{\varepsilon,\gamma}_{\kappa'\chi'}$ to (4.52) and (4.53) yields

$$\partial_x u_2' = \tfrac{1}{\varepsilon} P^{\varepsilon,\gamma}_{h^\varepsilon} u_2' + f_2' \,, \tag{4.55}$$

$$\partial_x v_2' = \tfrac{1}{\varepsilon} P^{\varepsilon,\gamma}_{\pi^\varepsilon} v_2' + g_2' \,, \tag{4.56}$$

where

$$\|f_2'\|_0 \le \|f_2\|_0 + \|u_2\|_0 \,, \qquad \|g_2'\|_0 \le \|g_2\|_0 + \|v_2\|_0 \,. \tag{4.57}$$

From (4.47), it follows that

$$\frac{1}{\varepsilon} P^{\varepsilon,\gamma}_{h^\varepsilon} u_2' = P^{\varepsilon,\gamma}_{h_0^\varepsilon} \partial_t u_2' + \sum_{j=1}^{d-1} P^{\varepsilon,\gamma}_{h_j^\varepsilon} \partial_j u_2' + \gamma P^{\varepsilon,\gamma}_{h_d^\varepsilon} u_2' \,.$$

With Proposition 3.19, this implies that

$$\|\frac{1}{\varepsilon} P^{\varepsilon,\gamma}_{h^\varepsilon} u_2' - T^\gamma_{\mathrm{h}^\varepsilon} u_2'\|_0 \lesssim \|u_2'\|_0 \,. \tag{4.58}$$

Therefore, we can replace equation (4.55) by

$$\partial_x u_2' = T^\gamma_{\mathrm{h}^\varepsilon} u_2' + \widetilde{f}_2' \,, \tag{4.59}$$

where $\widetilde{f}_2'$ satisfies $\|\widetilde{f}_2'\|_0 \lesssim \|f_2\|_0 + \|u_2\|_0$.

Introduce $\Sigma_1 = \mathrm{Re}\, T^\gamma_{\mathrm{s}_1^\varepsilon}$ and $\Sigma_2 = \mathrm{Re}\, P^{\varepsilon,\gamma}_{\sigma_2^\varepsilon}$. We consider the symmetrizers

$$S_k = \gamma \Sigma_k - \sum_{j=0}^{d-1} \varepsilon \partial_j \Sigma_k \partial_j \,, \qquad \text{for } k = 1, 2 \,.$$

We start from the identities

$$\begin{aligned} &\big(\!\big(S_1(0) u_2'(0), u_2'(0) \big)\!\big) + \mathrm{Re}\, \big(\!\big(S_1 T^\gamma_{\mathrm{h}^\varepsilon} u_2', u_2' \big)\!\big) \\ &\qquad = -\, \big(\!\big([\partial_x, S_1] u_2', u_2' \big)\!\big) - 2\mathrm{Re}\, \big(\!\big(\widetilde{f}_2', S_1 u_2' \big)\!\big) \,, \end{aligned} \tag{4.60}$$

$$\begin{aligned} &\big(\!\big(S_2(0) v_2'(0), v_2'(0) \big)\!\big) + \frac{1}{\varepsilon} \mathrm{Re}\, \big(\!\big(S_2 P^{\varepsilon,\gamma}_{e^\varepsilon} v_2', v_2') \big)\!\big) \\ &\qquad = -\, \big(\!\big([\partial_x S_2] v_2', v_2') \big)\!\big) - 2\mathrm{Re}\, \big(\!\big(S_2 g_2', v_2' \big)\!\big) \,. \end{aligned} \tag{4.61}$$

We now estimate the different terms.

a) The right hand side of (4.60). By Proposition 3.6, since $\partial_x \mathrm{s}^\varepsilon$ is bounded in $\Gamma^0_{0,0}$ $[\partial_x, T^\gamma_{\mathrm{s}^\varepsilon_1}]$ and therefore its adjoint and hence $[\partial_x, \Sigma_1]$, are uniformly bounded in L^2. Thus

$$\big|\big(\!\big((\partial_x S_1)u'_2, u'2\big)\!\big)\big| \lesssim \gamma \|u'_2\|_0^2 + \varepsilon \|\nabla_{\tilde y} u'_2\|_0^2 \lesssim \|u_2\|^2_{0,(\varphi)},$$

where the second inequality follows from (4.54).

Similarly, Σ_1 is of order zero in the sense of Proposition 3.4 and

$$\begin{aligned}\big|\big(\!\big(\widetilde{f'_2}, S_1 u'_2\big)\!\big)\big| \;\leq\; \|\widetilde{f'_2}\|_0 \, \|S_1 u'_2\|_0 &\lesssim \|f'_2\|_0 \left(\gamma \|u'_2\|_0 + \varepsilon \|\nabla^2_{\tilde y} u'_2\|_0\right) \\ &\lesssim \|\widetilde{f'_2}\|_0 \, \|u'_2\|_{0,(\varphi^2)} \,.\end{aligned}$$

b) The right hand side of (4.61). Since $\partial_x \sigma^\varepsilon_2$ is bounded in $\mathrm{P}\Gamma^0_{0,0}$, the commutators $[\partial_x, \Sigma_2]$ are uniformly bounded in L^2 and

$$\big|\big(\!\big((\partial_x S_2) v'_2, v'_2\big)\!\big)\big| \lesssim \gamma \|v'_2\|_0^2 + \varepsilon \|\nabla_{\tilde y} v_2\|_0^2 \lesssim \|v'\|^2_{0,(\varphi)} \,.$$

For the second term, we give a symmetric role to g_2 and v_2, using that Σ_2 is uniformly bounded in L^2:

$$\begin{aligned}\big|\big(\!\big((S_2 g'_2, v'_2)\big)\!\big)\big| &\lesssim \gamma \|g'_2\|_0 \, \|v'_2\|_0 + \sum_j \varepsilon \|\partial_j g'_2\|_0 \, \|\partial_j v'_2\|_0 \\ &\lesssim \|g'_2\|_{0,(\varphi)} \, \|v'_2\|_{0,(\varphi)} \,.\end{aligned}$$

c) We apply Gardings' inequality to the second term in the left hand side of (4.60). Since, s^ε is bounded in $\Gamma^0_{1,0}$, $(T^\gamma_{\mathrm{s}^\varepsilon})^* - T^\gamma_{\mathrm{s}^\varepsilon})$ is of order -1 by Proposition adjoint. Thus $\Sigma_1 - T^\gamma_{\mathrm{s}^\varepsilon}$ is also of order -1. Together with the symbolic calculus in Proposition 3.5, this implies that

$$S_1 T^\gamma_{\mathrm{h}^\varepsilon} = \gamma \left(T^\gamma_{\mathrm{s}^\varepsilon_1 \mathrm{h}^\varepsilon} + R_0\right) + \sum_j \varepsilon \partial_j \left(T^\gamma_{\mathrm{s}^\varepsilon_1 \mathrm{h}^\varepsilon} \partial_j + R^j_1\right),$$

where R_0 is of order zero and the R^j_1 are of order one in the sense of section 3.1. With notations as in Lemma 4.11, we have

$$\operatorname{Re} T^\gamma_{\mathrm{s}^\varepsilon_1 \mathrm{h}^\varepsilon} = \sum (T^\gamma_{\mathrm{v}^\varepsilon_l})^* T_{\mathrm{k}^\varepsilon_l} T^\gamma_{\mathrm{v}^\varepsilon_l} + R_0$$

with R_0 of order 0. Thus,

$$\begin{aligned}&\big(\!\big(S_1 T^\gamma_{\mathrm{h}^\varepsilon} u'_2, u'_1\big)\!\big) - \sum_l \gamma \big(\!\big(T_{\mathrm{k}^\varepsilon_l} T^\gamma_{\mathrm{v}^\varepsilon_l} u'_2, T^\gamma_{\mathrm{v}^\varepsilon_l} u'_2\big)\!\big) - \sum_{l,j} \varepsilon \big(\!\big(T_{\mathrm{k}^\varepsilon_l} T^\gamma_{\mathrm{v}^\varepsilon_l} \partial_j u'_2, T^\gamma_{\mathrm{v}^\varepsilon_l} \partial_j u'_2\big)\!\big) \\ &\qquad\qquad = O(\gamma \|u'_2\|_0^2 + \varepsilon \|\partial_{\tilde y} u'_2\|_0^2) = O(\|u'_2\|^2_{0,(\varphi)}) \,.\end{aligned}$$

Next, $T^\gamma_{\mathrm{v}^\varepsilon_l} T^\gamma_{\kappa'\chi'} - T^\gamma_{\mathrm{v}^\varepsilon_l \kappa'\chi'}$ is of order -1. Since k^ε_l is of order 1, in the formula above we can replace $T^\gamma_{\mathrm{v}^\varepsilon_l} u'_2$ and $T^\gamma_{\mathrm{v}^\varepsilon_l} \partial_j u'_2$ by $T^\gamma_{\mathrm{v}^\varepsilon_l \kappa'\chi'} u_2$ and $T^\gamma_{\mathrm{v}^\varepsilon_l \kappa'\chi'} \partial_j u_2$ respectively, at the price of an error of order $O(\|u_2\|^2_{0,(\varphi)})$.

We estimate the scalar products $\big(\!\big(T_{\mathrm{k}^\varepsilon_l} T^\gamma_{\mathrm{v}^\varepsilon_l} u, T^\gamma_{\mathrm{v}^\varepsilon_l} u\big)\!\big)$ using Garding's inequality. According to the block structure of k^ε_l, there is a decomposition

$$\mathrm{v}^\varepsilon_l = \begin{bmatrix} \mathrm{v}^\varepsilon_{1,l} \\ \vdots \\ \mathrm{v}^\varepsilon_{q,l} \end{bmatrix}$$

so that

$$\big(\!\big(T_{\mathrm{k}_l^\varepsilon} T^\gamma_{\mathrm{v}_l^\varepsilon} u, T^\gamma_{\mathrm{v}_l^\varepsilon} u\big)\!\big) = \sum_{j=1}^{q} \big(\!\big(T_{\mathrm{b}_j^\varepsilon} T^\gamma_{\mathrm{v}_{j,l}^\varepsilon} u T^\gamma_{\mathrm{v}_{j,l}^\varepsilon} u\big)\!\big)$$

When b_j is elliptic on $\omega_1 \times \varepsilon^{-1}\mathrm{C}_1$ and $w \in \Gamma^0_{1,0}$ is supported in this set, Garding's inequality implies that

$$\big(\!\big(T_{\mathrm{b}_j^\varepsilon} T^\gamma_w u, T^\gamma_w u\big)\!\big) \geq c\|T^\gamma_w u\|^2_{0,\frac{1}{2},\gamma} - C\|u\|^2_{0,-\frac{1}{2},\gamma}$$

with norms as in (3.6) (see Proposition 3.7 and Remark 3.8).

When $\mathrm{b}_j = \gamma \mathrm{b}_{j,0} + \varepsilon \mathrm{b}_{j,2}$ with $\mathrm{b}_{j,0} \in \Gamma^0_{1,0}$ elliptic and $\mathrm{b}_{j,2} \in \Gamma^2_{1,0}$ elliptic, then Garding's inequality implies that

$$\big(\!\big(T_{\mathrm{b}_j^\varepsilon} T^\gamma_w u, T^\gamma_w u\big)\!\big) \geq c\,(\gamma\|T^\gamma_w u\|^2_0 + \varepsilon\|T^\gamma_w u\|^2_{0,1,\gamma}) - C(\gamma\,\|u\|^2_{0,-1,\gamma} + \varepsilon\|u\|^2_0).$$

On the spectrum of $T^\gamma_w u$, we have $|\zeta| \geq \gamma + \varepsilon|\zeta|^2$; thus in both cases, we have

$$\big(\!\big(T_{\mathrm{b}_j^\varepsilon} T^\gamma_w u, T^\gamma_w u\big)\!\big) \geq c(\gamma\|T^\gamma_w u\|^2_0 + \varepsilon\|T^\gamma_w u\|^2_{0,1,\gamma}) - C\gamma^{-1}\|u\|^2_0\,.$$

We apply these estimates to $u = u_2$ or $u = \partial_{\tilde{y}} u_2$ and $w = \mathrm{v}^\varepsilon_{j,l}\kappa'\chi'$. Adding up, and using (4.54) for the remainders, we obtain

$$\begin{aligned}\operatorname{Re}\big(\!\big(S_1 T^\gamma_{\mathrm{h}^\varepsilon} u_2', u_2'\big)\!\big) \geq c \sum_l \gamma(\gamma\|T^\gamma_{\mathrm{v}_l^\varepsilon\kappa'\chi'} u_2\|^2_0 + \varepsilon\|T^\gamma_{\mathrm{v}_l^\varepsilon\kappa'\chi'} u_2\|^2_{0,1,\gamma})\\ + \sum_{l,j} \varepsilon(\gamma\|T^\gamma_{\mathrm{v}_l^\varepsilon\kappa'\chi'}\partial_j u_2\|^2_{0,} + \varepsilon\|T^\gamma_{\mathrm{v}_l^\varepsilon\kappa'\chi'}\partial_j u_2\|^2_{0,1,\gamma}) - C\|u_2\|^2_{0,(\varphi)}\,.\end{aligned}$$

Next we use the ellipticity of $\sum(\mathrm{v}_l)^*\mathrm{v}_l$ on the support of $\kappa'\chi'$ which implies that there are bounded families of symbols $\widetilde{\mathrm{v}}_l^\varepsilon$ in $\Gamma^0_{1,0}$ such that

$$\kappa'\chi' = \sum_l \widetilde{\mathrm{v}}_l^\varepsilon \mathrm{v}_l^\varepsilon \kappa'\chi'\,.$$

Thus

$$\|T^\gamma_{\kappa'\chi'} u\|^2_0 \lesssim \sum \|T^\gamma_{\mathrm{v}_l^\varepsilon\kappa'\chi'} u\|^2_0 + \|u\|^2_{0,-1,\gamma}\,.$$

Adding up all the estimates, and using (4.54), we have proved that

$$\operatorname{Re}\big(\!\big(S_1 T^\gamma_{\mathrm{h}^\varepsilon} u_2', u_2'\big)\!\big) \geq c'\|u_2'\|^2_{0,(\varphi^2)} - C'\|u_2\|^2_{0,(\varphi)}.$$

d) The second term in the left hand side of (4.61). The analysis of this *elliptic* term is quite similar to the analysis made for bounded frequencies in Proposition 4.10. The symbolic calculus implies that

$$\Sigma_2 P^{\varepsilon,\gamma}_{\pi^\varepsilon} = P^{\varepsilon,\gamma}_{\sigma_2^\varepsilon\pi^\varepsilon} + \varepsilon R$$

with R of order 0 (in the calculus of section 3.2). Moreover $[\partial_{\tilde{y}}, P^{\varepsilon,\gamma}_{\pi^\varepsilon}] = P^{\varepsilon,\gamma}_{\partial_{\tilde{y}} e^\varepsilon}$ is also of order zero.

$$S_2 P^{\varepsilon,\gamma}_{\pi^\varepsilon} = \gamma\, P^{\varepsilon,\gamma}_{\sigma_2^\varepsilon\pi^\varepsilon} + \varepsilon\gamma R_0 + \sum_j \varepsilon\partial_j\big(P^{\varepsilon,\gamma}_{\sigma_2^\varepsilon\pi^\varepsilon}\partial_j + R_j' + \varepsilon R_j''\partial_j\big)\,,$$

where the remainders R are of order 0. By Garding's inequality (cf Proposition 3.15 and Remark 3.16),

$$\begin{aligned}\gamma\operatorname{Re}\big(\!\big(P^{\varepsilon,\gamma}_{\sigma_2^\varepsilon\pi^\varepsilon} v_2', v_2'\big)\!\big) &\geq c\gamma\|v_2'\|^2_0 - C\varepsilon^2\|v_2\|^2_0\,,\\ \varepsilon\operatorname{Re}\big(\!\big(P^{\varepsilon,\gamma}_{\sigma_2^\varepsilon\pi^\varepsilon}\partial_{\tilde{y}} v_2', \partial_{\tilde{y}} v_2'\big)\!\big) &\geq c\varepsilon\|\partial_{\tilde{y}} v_2'\|^2_0 - C\varepsilon^3\,\|\partial_{\tilde{y}} v_2\|^2_0\,.\end{aligned}$$

Since $\varepsilon\gamma$ and $\varepsilon\partial_{\tilde{y}}$ are bounded in our domain of analysis and using also (4.54), we conclude that

$$\frac{1}{\varepsilon}\mathrm{Re}\,\big(\!\big(S_2 P^{\varepsilon,\gamma}_{\pi^\varepsilon} v_2', v_2'\big)\!\big) \geq \frac{c}{\varepsilon}\|v_2\|^2_{0,(\varphi)} - C\|v_2\|^2_0 .$$

e) The boundary terms. Since $\mathrm{a}^\varepsilon := \mathrm{s}^\varepsilon + C_0(\mathrm{g}^\varepsilon)^*\mathrm{g}^\varepsilon$ is definite positive on $\omega_1 \times \varepsilon^{-1}\mathrm{C}_1$, Proposition 3.7 implies that

$$c\big|U_2'(0)\big|^2_0 \leq \big(\!\big(T^\gamma_{\mathrm{a}^\varepsilon} U_2'(0), U_2'(0)\big)\!\big) + C\big|U_2(0)\big|^2_{0,-1,\gamma}$$

Next, we use again Proposition 3.19 and write

$$P^{\varepsilon,\gamma}_{\sigma_2^\varepsilon(0)} = T^\gamma_{\mathrm{s}_2^\varepsilon(0)} + R_{-1}\,, \quad P^{\varepsilon,\gamma}_{\widetilde{\Gamma}^\varepsilon} = T^\gamma_{\mathrm{g}^\varepsilon} + R_{-1}$$

where the remainders R_{-1} are of order -1 in the sense of (3.27). The symbolic calculus yields $\Sigma_2(0) = T^\gamma_{\mathrm{s}_2^\varepsilon(0)} + R_{-1}$ and thus

$$\Sigma(0) := \begin{pmatrix} \Sigma_1(0) & 0 \\ 0 & \Sigma_2(0) \end{pmatrix} = T^\gamma_{\mathrm{s}^\varepsilon} + R_{-1}$$

and

$$T^\gamma_{a^\varepsilon} = \Sigma(0) + C(P^{\varepsilon,\gamma}_{\widetilde{\Gamma}^\varepsilon})^* P^{\varepsilon,\gamma}_{\widetilde{\Gamma}^\varepsilon} + R_{-1}\,.$$

Therefore,

$$\big|U_2'(0)\big|^2_0 \lesssim \big(\!\big(\Sigma(0)U_2'(0), U_2'(0)\big)\!\big) + \big|P^{\varepsilon,\gamma}_{\widetilde{\Gamma}^\varepsilon} U_2'(0)\big|^2_0 + \big|U_2(0)\big|^2_{0,-1,\gamma}\,.$$

There is a similar estimate for $\partial_{\tilde{y}} U_2'$. Using again (4.54), yields

$$\big|U_2(0)\big|^2_{0,(\varphi)} \lesssim \big(\!\big(S(0)U_2(0), U(0)\big)\!\big) + \big|P^{\varepsilon,\gamma}_{\widetilde{\Gamma}^\varepsilon} U_2(0)\big|^2_{0,(\varphi)} + \big|U_2(0)\big|^2_0\,.$$

f) Adding (4.60) and (4.61) and using the estimates above, one obtains

$$\begin{aligned}\|u_2'\|^2_{0,(\varphi^2)} + \frac{1}{\varepsilon}\|v_2'\|^2_{0,(\varphi)} + \big|U_2'(0)\big|^2_{0,(\varphi)} \lesssim \\ \|\widetilde{f}_2\|_0\,\|u_2'\|_{0,(\varphi^2)} + \|g_2'\|_{0,(\varphi)}\,\|v_2'\|_{0,(\varphi)} + \big|P^{\varepsilon,\gamma}_{\widetilde{\Gamma}^\varepsilon} U_2(0)\big|^2_{0,(\varphi)} \\ + \|u_2\|^2_{0,(\varphi)} + \|v_2\|^2_0 + \big|U_2(0)\big|^2_0\end{aligned}$$

and the proposition follows. □

PROPOSITION 4.13. *There are C and γ_0 such that for all $u \in C_0^\infty(\overline{\omega})$, $\varepsilon \in]0,1$ and $\gamma \geq \gamma_0$, the function $U_2 = P^{\varepsilon,\gamma}_{w^\varepsilon} U$ satisfies*

$$\begin{aligned}\|u_2\|_{m,(\varphi^2)} + \frac{1}{\sqrt{\varepsilon}}\|v_2\|_{m,(\varphi)} + |U_2(0)|_{m,(\varphi)} \leq C\Big(\|f_2\|_m \\ + \sqrt{\varepsilon}\|g_2\|_{m,(\varphi)} + |P^{\varepsilon,\gamma}_{\widetilde{\Gamma}^\varepsilon} U_2(0)|_{m,(\varphi)} + \|U\|_m + |U(0)|_m\Big)\,.\end{aligned}$$

PROOF. We differentiate the equations (4.52) and (4.53) with respect to Z^α, $|\alpha| \leq m$.

a) Equation for $Z^\alpha u_2$.

$$[Z^\alpha, P^{\varepsilon,\gamma}_{h^\varepsilon}] = \sum_{\beta<\alpha} \binom{\alpha}{\beta} P^{\varepsilon,\gamma}_{Z^{\alpha-\beta}h^\varepsilon} Z^\beta\,.$$

Using (4.47), one obtains

$$P^{\varepsilon,\gamma}_{Z^{\alpha-\beta}h^\varepsilon} = \varepsilon\Big(\gamma P^{\varepsilon,\gamma}_{Z^{\alpha-\beta}h^\varepsilon_d} + i\sum_{j=0}^{d-1} P^{\varepsilon,\gamma}_{Z^{\alpha-\beta}h^\varepsilon_j}\partial_{\tilde{y}_j}\Big)\,.$$

Since h^ε and thus the h^ε_j are bounded in $\mathrm{P}\Gamma^0_{1,m}$, the $Z^{\alpha-\beta}h^\varepsilon_j$ are bounded in $\mathrm{P}\Gamma^0_{1,0}$ and the operators $P^{\varepsilon,\gamma}_{Z^{\alpha-\beta}h^\varepsilon_j}$ are uniformly bounded in L^2. Since $\beta < \alpha$, we can incorporate the extra $\partial_{\tilde{y}_j}$ to Z^β to obtain a conormal derivative $Z^{\beta'}$ with $|\beta'| \le |\alpha|$. Therefore

$$\gamma^{m-|\alpha|}\, \|[Z^\alpha, P^{\varepsilon,\gamma}_{h^\varepsilon}]u_2\|_0 \lesssim \varepsilon \, \|u_2\|_m \, .$$

The commutator $[Z^\beta, \partial_x]u_2$ is a sum of terms $Z^\beta\partial_x u_2 = Z^\beta f_2 + \varepsilon^{-1}Z^\beta P^{\varepsilon,\gamma}_{h^\varepsilon}u_2$ with $\beta < \alpha$. We repeat the same argument to commute Z^β to $P^{\varepsilon,\gamma}_{h^\varepsilon}$, using that $|\beta| < m$. This shows that

$$\begin{aligned} &\partial_x Z^\alpha u_2 = \frac{1}{\varepsilon} P^{\varepsilon,\gamma}_{h^\varepsilon} Z^\alpha u_2 + f^\alpha \, , \\ &\gamma^{m-|\alpha|}\|f^\alpha\|_0 \lesssim \|f\|_m + \|u_2\|_m \, . \end{aligned} \tag{4.62}$$

b) Equation for $Z^\alpha v_2$. We note that for $w \in \mathrm{P}\Gamma^0_{1,0}$ and v such that $\varepsilon\zeta$ is bounded on its spectrum one has

$$\|P^{\varepsilon,\gamma}_w v\|_{0,(\varphi)} \lesssim \|v\|_{0,(\varphi)} \, .$$

Indeed, using Proposition 3.4 and (4.54), there holds

$$\begin{aligned} \|P^{\varepsilon,\gamma}_w v\|_{0,(\varphi)} &\le \sqrt{\gamma}\|P^{\varepsilon,\gamma}_w v\|_0 + \sqrt{\varepsilon}\|\partial_{\tilde{y}} P^{\varepsilon,\gamma}_w v\|_0 \\ &\lesssim \sqrt{\gamma}\|v\|_0 + \sqrt{\varepsilon}\|\partial_{\tilde{y}} v\|_0 \lesssim \|v\|_{0,(\varphi)} \, . \end{aligned}$$

Thus

$$\|[Z^\alpha, P^{\varepsilon,\gamma}_{e^\varepsilon}]v_2\|_{0,(\varphi)} \lesssim \sum_{\beta<\alpha} \|Z^\beta v_2\|_{0,(\varphi)}$$

and

$$\gamma^{m-|\alpha|}\, \|[Z^\alpha, P^{\varepsilon,\gamma}_{e^\varepsilon}]v_2\|_{0,(\varphi)} \lesssim \frac{1}{\gamma}\|v_2\|_{m,(\varphi)} \, .$$

For future use, we note that we have proved that

$$\|P^{\varepsilon,\gamma}_w v\|_{m,(\varphi)} \lesssim \|v\|_{m,(\varphi)} \tag{4.63}$$

if $\varepsilon\zeta$ is bounded on the spectrum of v.

The commutator $[Z^\beta, \partial_x]v_2$ is a sum of terms $Z^\beta g_2 + \varepsilon^{-1}Z^\beta P^{\varepsilon,\gamma}_{e^\varepsilon}u_2$ with $\beta < \alpha$ and, finally, we see that

$$\begin{aligned} &\partial_x Z^\alpha v_2 = \frac{1}{\varepsilon} P^{\varepsilon,\gamma}_{e^\varepsilon} Z^\alpha v_2 + g^\alpha \, , \\ &\gamma^{m-|\alpha|}\|g^\alpha\|_0 \lesssim \|g\|_{m,(\varphi)} + \frac{1}{\varepsilon\gamma}\|v_2\|_{m,(\varphi)} \, . \end{aligned} \tag{4.64}$$

c) The traces. One has $Z^\alpha U_2(0) = 0$, except when $Z^\alpha = \partial^\alpha_{\tilde{y}}$ contains no $x\partial_x$ derivatives. In this case, repeating the analysis in b), one obtains

$$\gamma^{m-|\alpha|}\, \big|[Z^\alpha, P^{\varepsilon,\gamma}_{\tilde{\Gamma}^\varepsilon}]U_2(0)\big|_{0,(\varphi)} \lesssim \frac{1}{\gamma}\, \big|U_2(0)\big|_{m,(\varphi)} \, .$$

d) We apply Proposition 4.12 for each $Z^\alpha U_2$ and $|\alpha| \le m$. together with the commutator estimates above, this yields

$$\begin{aligned} \gamma^{m-|\alpha|}\Big(\|P^{\varepsilon,\gamma}_{\kappa'\chi'} Z^\alpha u_2\|_{0,(\varphi^2)} &+ \frac{1}{\sqrt{\varepsilon}}\|P^{\varepsilon,\gamma}_{\kappa'\chi'} Z^\alpha v_2\|_{0,(\varphi)} + |P^{\varepsilon,\gamma}_{\kappa'\chi'} Z^\alpha U_2(0)|_{0,(\varphi)}\Big) \\ &\lesssim \|f_2\|_m + \sqrt{\varepsilon}\|g_2\|_{m,(\varphi)} + |P^{\varepsilon,\gamma}_{\tilde{\Gamma}^\varepsilon} U_2(0)|_{m,(\varphi)} + \|u_2\|_{m,(\varphi)} \\ &\quad + \|v_2\|_m + |U_2(0)|_m + \frac{1}{\gamma\sqrt{\varepsilon}}\|v_2\|_{m,(\varphi)} + \frac{1}{\gamma}|U_2(0)|_{m,(\varphi)} \, . \end{aligned}$$

We now argue as in Lemma 4.8, using that $U_2 = P^{\varepsilon,\gamma}_{w^\varepsilon} U$ and $\kappa_1'\chi' = 1$ on the support of w^ε. Using (4.54), we see that

$$\begin{aligned}
\|u_2\|_{m,(\varphi^2)} &\lesssim \sum_{|\alpha|\le m} \gamma^{m-|\alpha|} \|P^{\varepsilon,\gamma}_{\kappa'\chi'} Z^\alpha u_2\|_{0,(\varphi^2)} + \|U\|_m\,,\\
\|v_2\|_{m,(\varphi)} &\lesssim \sum_{|\alpha|\le m} \gamma^{m-|\alpha|} \|P^{\varepsilon,\gamma}_{\kappa'\chi'} Z^\alpha v_2\|_{0,(\varphi)} + \sqrt{\varepsilon}\|U\|_m\,,\\
|U_2(0)|_{m,(\varphi)} &\lesssim \sum_{|\alpha|\le m} \gamma^{m-|\alpha|} |P^{\varepsilon,\gamma}_{\kappa'\chi'} Z^\alpha U_2(0)|_{0,(\varphi^2)} + \sqrt{\varepsilon}\|U(0)\|_m\,.
\end{aligned}$$

Therefore,

$$\begin{aligned}
\|u_2\|_{m,(\varphi^2)} + \frac{1}{\sqrt{\varepsilon}}\|v_2\|_{m,(\varphi)} + |U_2(0)|_{m,(\varphi)} &\lesssim \|f_2\|_m \\
&+\sqrt{\varepsilon}\|g_2\|_{m,(\varphi)} + |P^{\varepsilon,\gamma}_{\widetilde{\Gamma}^\varepsilon} U_2(0)|_{m,(\varphi)} + \|U\|_m + |U(0)|_m \\
&\quad + \|u_2\|_{m,(\varphi)} + \frac{1}{\gamma\sqrt{\varepsilon}}\|v_2\|_{m,(\varphi)} + \frac{1}{\gamma}|U_2(0)|_{m,(\varphi)}\,.
\end{aligned}$$

When γ is large enough, the last three terms the right hand side can be absorbed from the right to the left and the proposition follows. □

Having proved the estimates for U_2, we now prove the estimates for $U_1 = P_{\kappa\chi} U$.

PROPOSITION 4.14. *With notations as above, there are C and $\gamma_0 \ge 1$ such that the estimate* (4.24) *is satisfied by $P_{\kappa\chi}U$. for all $\gamma \ge \gamma_0$, all $\varepsilon \in]0,1]$ and all $u \in C_0^\infty(\overline{\omega})$.*

PROOF. We fix the neighborhoods and cut-off functions so that Proposition 4.13 applies. Before estimating U_1 we need some preparation.

a) First, we remark with (4.54) that

$$\|v_2\|_{m,(\varphi)^2} \lesssim \frac{1}{\sqrt{\varepsilon}}\|v_2\|_{m,(\varphi)}\,, \tag{4.65}$$

$$\sqrt{\varepsilon}\|g_2\|_{m,(\varphi)} \lesssim \|g_2\|_m\,. \tag{4.66}$$

Next, we note that for $\varepsilon|\zeta|$ bounded, $\varepsilon|\zeta| \le C\sqrt{\varepsilon}\varphi$ and thus $\varepsilon|\zeta|\varphi/\sqrt{\varepsilon} \le C\varphi^2$. This implies that

$$\frac{1}{\sqrt{\varepsilon}}\|\varepsilon\partial_{\tilde y} u_2\|_{m,(\varphi)} + \frac{1}{\sqrt{\varepsilon}}\|\varepsilon\gamma u_2\|_{m,(\varphi)} \lesssim \|u_2\|_{m,(\varphi^2)}\,.$$

If $w \in \mathbb{P}\Gamma^0_{1,m}$ has compact support in ζ and vanishes at $\zeta = 0$, one can write $w = \sum w_j i\tilde\eta_j + \gamma w_d$ with $w_j \in \mathbb{P}\Gamma^0_{1,m}$ and $P^{\varepsilon,\gamma}_w = \sum P^{\varepsilon,\gamma}_{w_j}\varepsilon\partial_{\tilde y_j} + P^{\varepsilon,\gamma}_{w_d}\varepsilon\gamma$. Using (4.63) for the action of the $P^{\varepsilon,\gamma}_{w_j}$, we get

$$\frac{1}{\sqrt{\varepsilon}}\|P^{\varepsilon,\gamma}_w u_2\|_{m,(\varphi)} \lesssim \|u_2\|_{m,(\varphi^2)}\,. \tag{4.67}$$

Moreover, $\varepsilon\varphi/\sqrt{\varepsilon}$ is bounded when $\varepsilon|\zeta|$ is bounded and thus

$$\frac{1}{\sqrt{\varepsilon}}\|\varepsilon f_2\|_{m,(\varphi)} \lesssim \|f_2\|_m\,.$$

With equation (4.52), and applying (4.67) to $w = h^\varepsilon$ which vanishes at $\zeta = 0$, we conclude that

$$\frac{1}{\sqrt{\varepsilon}}\|\varepsilon\partial_x u_2\|_{m,(\varphi)} \lesssim \|u_2\|_{m,(\varphi^2)} + \|f_2\|_m . \tag{4.68}$$

Writing

$$u_2(x) = u_2(0) + \int_0^x \partial_x u_2(x')dx' ,$$

one obtains that for $\theta > 0$,

$$\|e^{-\theta x/\varepsilon} u_2\|_0 \leq C\left(\sqrt{\varepsilon}\big|u_2(0)\big|_0 + \|\varepsilon\partial_x u_2\|_0\right).$$

This also holds for the tangential Fourier transform of u_2, and extends to any weighted space with weight depending only on ζ, in particular to the norms $(0,(\varphi))$. Next we commute with Z^α. Commuting with ∂_y^α is trivial. Next we write

$$e^{-\theta x/\varepsilon} x\partial_x u_2 = \Phi(x/\varepsilon)\varepsilon\partial_x u_2 \quad \text{with } \Phi(z) = ze^{-\theta z},$$

yielding the estimate

$$\|e^{-\theta x/\varepsilon} u_2\|_{m,(\varphi)} \leq C\left(\sqrt{\varepsilon}\big|u_2(0)\big|_{m,(\varphi)} + \|\varepsilon\partial_x u_2\|_{m,(\varphi)}\right). \tag{4.69}$$

With (4.68), we conclude that

$$\frac{1}{\sqrt{\varepsilon}}\|e^{-\theta x/\varepsilon} u_2\|_{m,(\varphi)} \lesssim \|u_2\|_{m,(\varphi^2)} + \big|u_2(0)\big|_{m,(\varphi)} + \|f_2\|_m . \tag{4.70}$$

b) We now proceed to the proof of (4.24). The symbol w_{-1}^ε satisfies

$$w_{-1}^\varepsilon w^\varepsilon = \kappa(t,y,x)\,\chi(\zeta)\mathrm{Id}.$$

Therefore

$$U_1 = P_{\kappa\chi}^{\varepsilon,\gamma} U = P_{w_{-1}^\varepsilon}^{\varepsilon,\gamma} U_2 + \varepsilon RU , \tag{4.71}$$

where R is of order zero. Moreover, $\varepsilon\zeta$ is bounded on the spectrum of the first two terms and thus on the spectrum of RU. Therefore

$$\|\varepsilon RU\|_{m,(\varphi^2)} \lesssim \frac{1}{\sqrt{\varepsilon}}\|\varepsilon RU\|_{m,(\varphi)} \lesssim \|RU\|_m \lesssim \|U\|_m . \tag{4.72}$$

From (4.71) and the symbolic calculus, we get that

$$u_1 = P_{w_\flat^\varepsilon}^{\varepsilon,\gamma} U_2 + \varepsilon r ,$$

where $w_\flat^\varepsilon$ is the upper $N \times 2N$ block of w_{-1}^ε and εr satisfies estimates analogous to (4.72). As for (4.63), we note that

$$\|P_{w_\flat^\varepsilon}^{\varepsilon,\gamma} U_2\|_{m,(\varphi^2)} \lesssim \|U_2\|_{m,(\varphi^2)} . \tag{4.73}$$

Indeed, since $\varepsilon\zeta$ is bounded on the spectrum of U_2 and $P_{w_\flat^\varepsilon}^{\varepsilon,\gamma} U_2$, one has

$$\|P_{w_\flat^\varepsilon}^{\varepsilon,\gamma} U_2\|_{0,(\varphi^2)} \approx \gamma\|P_{w_\flat^\varepsilon}^{\varepsilon,\gamma} U_2\|_0 + \varepsilon^{-1}\|P_{|\zeta|^2}^{\varepsilon,\gamma} P_{w_\flat^\varepsilon}^{\varepsilon,\gamma} U_2\|_0 .$$

Since

$$P_{|\zeta|^2}^{\varepsilon,\gamma} P_{w_\flat^\varepsilon}^{\varepsilon,\gamma} = P_{w_\flat^\varepsilon}^{\varepsilon,\gamma} P_{|\zeta|^2}^{\varepsilon,\gamma} + \varepsilon R$$

with R of degree zero since $w_\flat^\varepsilon$ is bounded in $\mathrm{P}\Gamma_{1,0}^0$. Therefore,

$$\|P_{w_\flat^\varepsilon}^{\varepsilon,\gamma} U_2\|_{0,(\varphi^2)} \lesssim \gamma\|U_2\|_0 + \varepsilon^{-1}\|P_{|\zeta|^2}^{\varepsilon,\gamma} U_2\|_0 + \|U_2\|_0 \lesssim \|U_2\|_{0,(\varphi^2)} .$$

Commuting $P^{\varepsilon,\gamma}_{w^\varepsilon_\flat}$ and Z^α for $|\alpha| \le m$ and using that $Z^{\alpha-\beta} w^\varepsilon_\flat \in \mathrm{P}\Gamma^0_{1,0}$ yields (4.73). Together with (4.72) we get that

$$\|u_1\|_{m,(\varphi^2)} \lesssim \|U_2\|_{m,(\varphi)^2} + \|U\|_m . \tag{4.74}$$

Next we remember that the matrices $\mathcal{W}$ and $\mathcal{V}$ which enter in the definition of w^ε_{-1} satisfy

$$\mathcal{W}(z,p,\zeta) - \mathrm{Id} = e^{-\theta z}\mathcal{W}'(z,p,\zeta)\,, \quad \mathcal{V}(p,0) = \begin{pmatrix} \mathrm{Id} & * \\ 0 & \mathrm{Id} \end{pmatrix},$$

with $\mathcal{W}'$ uniformly bounded and smooth. This shows that

$$v_1 = P^{\varepsilon,\gamma}_{w^\varepsilon_2} v_2 + P^{\varepsilon,\gamma}_{w_1} u_2 + P^{\varepsilon,\gamma}_{w'}(e^{-\theta x/\varepsilon} u_2) + \varepsilon R U\,,$$

where w_1, w_2 and w' are symbols in $\mathrm{P}\Gamma^0_{1,m}$ such that $w_1(t,y,x,0) = 0$, and R is of degree zero as in (4.71). With (4.63), (4.67) and (4.70) we deduce that

$$\begin{aligned} \frac{1}{\sqrt{\varepsilon}}\|v_1\|_{m,(\varphi)} \lesssim\ & \frac{1}{\sqrt{\varepsilon}}\|v_2\|_{m,(\varphi)} + \|u_2\|_{m,(\varphi^2)} \\ & + \big|u_2(0)\big|_{m,(\varphi)} + \|f_2\|_m + \|U\|_m . \end{aligned} \tag{4.75}$$

Similarly, one has $\widetilde{\Gamma}^\varepsilon w^\varepsilon = \kappa\chi\Gamma$ and thus $P^{\varepsilon,\gamma}_{\kappa\chi} = P^{\varepsilon,\gamma}_{\widetilde{\Gamma}^\varepsilon} P^{\varepsilon,\gamma}_{w^\varepsilon} + \varepsilon R$ with R of order 0. Since $\Gamma U = u = 0$ at $x = 0$, one has $P^{\varepsilon,\gamma}_{\widetilde{\Gamma}^\varepsilon} U_2(0) = \varepsilon R U(0)$. Thus

$$\big|P^{\varepsilon,\gamma}_{\widetilde{\Gamma}^\varepsilon} U_2(0)\big|_{m,(\varphi)} \lesssim \big|U_2(0)\big|_m . \tag{4.76}$$

Collecting the various estimates, Proposition 4.14 implies that

$$\begin{aligned} \|u_1\|_{m,(\varphi^2)} + \frac{1}{\sqrt{\varepsilon}}\|v_1\|_{m,(\varphi)} + |v(0)|_{m,(\varphi)} \lesssim & \\ \|f_2\|_m + \|g_2\|_m + \|U\|_m + |U(0)|_m . & \end{aligned}$$

Together with the estimates (4.41) for $F_2 = (f_2, g_2)$, this implies (4.24) and the proposition is proved. □

4.6. Proof of Proposition 4.4. One first uses Propositions 4.7 and 4.14 to find a neighborhood ω and cut-off functions χ_0 and χ_R with χ_0 supported in a small neighborhood of the origin and $\chi_R = 1$ for large ζ. In the intermediate zone, we use Proposition 4.10 to find a finite covering, decreasing ω a finite number of times.

5. Linear stability

In this section, we prove the main stability estimates announced in Theorems 1.9 and 1.10. Note that they contain both conormal $\mathcal{H}^m$ estimates and L^∞ type estimates, which are crucial for the analysis of nonlinear stability in the next section. The basic conormal estimates for functions globally defined in time are stated in Theorem 5.1. By standard techniques, they imply local in time estimates and the well posedness of the linearized equation (Section 5.1). In Section 5.2, we show that the estimates of Theorem 5.1 are combinations of the estimates of Theorem 4.1 near the boundary and of estimates in the interior that are stated in Proposition 5.5 and proved in Section 5.3. Finally, the L^∞ estimates are proved in Section 5.4.

Consider functions u_0, b_0, and v^ε in $W^{m+2,\infty}([-T_0, T_0] \times \Omega)$, with v^ε satisfying (4.1). Introduce u_a as in (4.2), and consider the linearized equation (4.3) of (1.25) around u_a. As in section 4, we extend u_0, v^ε and thus u_a to $\mathbb{R} \times \Omega$, and we assume that the extended function are independent of t for $|t| \geq T_1 > T_0$. We first consider the linearized equation on $\mathbb{R} \times \Omega$:

$$
(5.1) \qquad \begin{cases} \mathcal{P}^\varepsilon u := \partial_t u + \sum_{j=1}^{d} A_j^\varepsilon \partial_j u - \varepsilon \sum_{j,k=1}^{d} B_{j,k}^\varepsilon \partial_{j,k}^2 u + \frac{1}{\varepsilon} E^\varepsilon u = f\,, \\ u_{|\mathbb{R}\times\partial\Omega} = 0\,. \end{cases}
$$

We introduce the space $\mathcal{H}_\gamma^m$ of functions $u \in e^{\gamma t} L^2(\mathbb{R} \times \Omega)$ such that $Z^I u \in e^{\gamma t} L^2(\mathbb{R} \times \Omega)$ for all sequence $I = (i_1, \ldots, i_k)$ of length $k = |I| \leq m$, where $Z^I = Z_{i_1} \cdots Z_{i_k}$. As usual, we agree that $Z^I = Id$ when $|I| = 0$. This space is equipped with the norm

$$
(5.2) \qquad \|u\|_{\mathcal{H}_\gamma^m} := \sum_{|I|\leq m} \gamma^{m-|I|}\, \|e^{-\gamma t}\, Z^I u\|_{L^2(\mathbb{R}\times\Omega)}, .
$$

The basic estimate is the following.

THEOREM 5.1. *There are $C > 0$ and γ_0 such that for all $\varepsilon \in]0, 1]$, all $\gamma \geq \gamma_0$ and all u in $C_0^\infty(\mathbb{R} \times \overline{\Omega})$ vanishing on $\mathbb{R} \times \partial\Omega$, one has*

$$
(5.3) \qquad \begin{aligned} \gamma\|u\|_{\mathcal{H}_\gamma^m} + \sqrt{\varepsilon\gamma}\|\nabla_x u\|_{\mathcal{H}_\gamma^m} + \sqrt{\varepsilon}\|\partial_t u\|_{\mathcal{H}_\gamma^m} &+ \varepsilon^{3/2}\|\nabla_x^2 u\|_{\mathcal{H}_\gamma^m} \\ &\leq C\, \|\mathcal{P}^\varepsilon u\|_{\mathcal{H}_\gamma^m}\,. \end{aligned}
$$

5.1. Proof of Theorem 1.10 part one, assuming Theorem 5.1. Denote by $\mathcal{H}_{\gamma,0}^m$ the space of functions $f \in \mathcal{H}^m$ which vanish for $t < 0$. Introduce next the space $\mathcal{K}_\gamma^m$ of functions $u \in \mathcal{H}_\gamma^m$ such that $\nabla_{t,x} u \in \mathcal{H}_\gamma^m$, $\nabla_x^2 u \in \mathcal{H}_\gamma^m$ and vanishing on the boundary $\mathbb{R} \times \partial\Omega$. Let $\mathcal{K}_{\gamma,0}^m$ denote the subspace of the $u \in \mathcal{K}_\gamma^m$ which vanish for $t < 0$. Note that for all $\varepsilon > 0$, $\mathcal{P}^\varepsilon$ is a bounded operator from $\mathcal{K}_\gamma^m$ to $\mathcal{H}_\gamma^m$ and from $\mathcal{K}_{\gamma,0}^m$ to $\mathcal{H}_{\gamma,0}^m$. We denote by $\mathcal{P}_\gamma^\varepsilon$ this mapping.

LEMMA 5.2. *For all $\gamma \geq \gamma_0$ and $\varepsilon \in]0, 1]$, P_γ^ε is an isomorphism from $\mathcal{K}_{\gamma,0}^m$ to $\mathcal{H}_{\gamma,0}^m$ and for all $u \in \mathcal{K}_{\gamma,0}^m$ the estimate* (5.3) *holds.*

Moreover, if $f \in \mathcal{H}_{\gamma,0}^m$ vanishes for $t > T_0$, then $u = (\mathcal{P}_\gamma^\varepsilon)^{-1} f$ also vanishes for $t < T_0$.

PROOF. By density, the estimate (5.3) extends to functions $u \in \mathcal{K}_\gamma^m$, and thus $\mathcal{P}_\gamma$ is injective for $\gamma \geq \gamma_0$.

Next, we use the classical theory of parabolic systems, for a given *fixed* $\varepsilon \in]0, 1]$. It implies that there is $\gamma(\varepsilon)$ such that for $\gamma \geq \gamma(\varepsilon)$ and $f \in \mathcal{H}_{\gamma,0}^m$ there is a unique solution u of (5.1) in $\mathcal{K}_{\gamma,0}^m$. This shows that $\mathcal{P}_\gamma^\varepsilon$ is an isomorphism from $\mathcal{K}_{\gamma,0}^m$ to $\mathcal{H}_{\gamma,0}^m$ for $\gamma \geq \gamma(\varepsilon)$. We show that (5.3) implies that it is an isomorphism for $\gamma \geq \gamma_0$. Indeed, let γ_* denote the infimum of the set of $\gamma \geq \gamma_0$ such that $\mathcal{P}_\gamma^\varepsilon$ is an isomorphism. For the given ε, the norms of $\mathcal{P}_\gamma^\varepsilon$ and, by (5.3), the norms of $(\mathcal{P}_\gamma^\varepsilon)^{-1}$ are bounded by uniform constants when $\gamma_* < \gamma \leq \gamma(\varepsilon)$. Therefore, there is $\delta_0 = \delta_0(\varepsilon) > 0$, independent of $\gamma > \gamma_*$ such that $P_\gamma^\varepsilon - \delta\mathrm{Id}$ is an isomorphism from $\mathcal{K}_{\gamma,0}^m$ to $\mathcal{H}_{\gamma,0}^m$ for $|\delta| \leq \delta_0$. Next we note that

$$
\mathcal{P}_{\gamma-\delta}^\varepsilon = e^{-\delta t}\, (P_\gamma^\varepsilon - \delta\mathrm{Id})\, e^{\delta t}\,.
$$

Therefore, if $\gamma > \gamma_* \geq \gamma_0$, $P_{\gamma'}$ is still an isomorphism for $\gamma' \in [\gamma - \delta_0, \gamma + \delta_0]$. This implies that $\gamma_* = \gamma_0$ and the first part of the lemma follows.

Suppose that f vanishes for $t < T_0$. Since $\mathcal{H}^m_{\gamma,0} \subset \mathcal{H}^m_{\gamma',0}$ for $\gamma' \geq \gamma$, we can use estimate (5.3) for $m = 0$ and $\gamma' \geq \gamma$. It implies that the L^2 norm of $e^{\gamma'(T-t)}u$ is bounded as γ' tends to infinity and thus $u = 0$ on $]-\infty, T_0[\times\Omega$. □

Consider now $f \in \mathcal{H}^m([-T_0, T_0] \times \Omega)$ which vanishes for $t < 0$. We can extend it to $t > T_0$, as a function $f^* \in \mathcal{H}^m(\mathbb{R} \times \Omega)$ which vanishes for $t > T_1$, for some given $T_1 > T_0$. Therefore, $f^* \in \mathcal{H}^m_{\gamma,0}$ for all $\gamma \geq 1$. Moreover, we can construct the extension so that

$$\|f^*\|_{\mathcal{H}^m_{\gamma_0}} \leq C\,\|f\|_{\mathcal{H}^m}\,, \tag{5.4}$$

where γ_0 is given by Theorem 5.1 and we use the notation $\|\cdot\|_{\mathcal{H}^m}$ for the norm in $\mathcal{H}^m([-T_0, T_0] \times \Omega)$ as in section 1.

Lemma 5.2 implies that the equation (5.1) with right hand side f^* as a solution $u^* \in \mathcal{K}^m_{\gamma_0,0}$. Therefore $u = u^*{}_{|\{t<T_0\}}$ is a solution of

$$\mathcal{P}^\varepsilon u = f\,, \quad u_{|[-T_0,T_0]\times\partial\Omega} = 0\,, \quad u_{|t<0} = 0\,.$$

The second part of the lemma implies that u is independent of the extension f^* of f. By standard uniqueness for parabolic systems, it is also independent of the extension of the coefficients.

For $t \in [-T_0, T_0]$ the weight $e^{\gamma_0 t}$ is bounded and thus, with notations as in section 1,

$$\|u\|_{\mathcal{H}^m} \leq C\,\|u^*\|_{\mathcal{H}^m_\gamma}\,. \tag{5.5}$$

There are similar estimates for the derivatives.

This proves the existence and uniqueness part in Theorems 1.9 and 1.10, and it is now clear that the L^2 and $\mathcal{H}^m$ estimates in these theorems follow from (5.3) (5.4) and (5.5).

5.2. Reduction to interior estimates. It remains to prove Theorem 5.1. As in section 4, we change the unknown u to $\widetilde{u} = e^{-\gamma t}u$ and the source term f to $\widetilde{f} = e^{-\gamma t}f$. In this case, the equation (5.1) reads

$$\begin{cases} \widetilde{\mathcal{P}}^\varepsilon \widetilde{u} := (\mathcal{P}^\varepsilon + \gamma)\widetilde{u} = \widetilde{f}\,, \\ \widetilde{u}_{|\,\mathbb{R}\times\partial\Omega} = 0\,. \end{cases} \tag{5.6}$$

We introduce the weighted norms

$$\|u\|_{\mathcal{H}^m,\gamma} := \sum_{|I|\leq m} \gamma^{m-|I|}\,\|Z^I u\|_{L^2(\mathbb{R}\times\Omega)}\,. \tag{5.7}$$

Since the commutators of Z^I with $e^{-\gamma t}$ are of the form $\gamma^{|J|}Z^{I-J}$ it is clear that

$$\frac{1}{C}\|e^{-\gamma t}u\|_{\mathcal{H}^m,\gamma} \leq \|u\|_{\mathcal{H}^m_\gamma} \leq C\|e^{-\gamma t}u\|_{\mathcal{H}^m,\gamma} \tag{5.8}$$

with C independent of $\gamma \geq 1$. Therefore, it is sufficient to prove that there are C and γ_0 such that for $\gamma \geq \gamma_0$, $\varepsilon \in]0,1]$, $\widetilde{u} \in C^\infty_0(\mathbb{R} \times \overline{\Omega})$ and $\widetilde{f}$ given by (5.6), one has

$$\gamma\|\widetilde{u}\|_{\mathcal{H}^m,\gamma} + \sqrt{\varepsilon\gamma}\|\nabla_x\widetilde{u}\|_{\mathcal{H}^m,\gamma} + \sqrt{\varepsilon}\|\partial_t\widetilde{u}\|_{\mathcal{H}^m,\gamma} + \varepsilon^{3/2}\|\nabla_x^2\widetilde{u}\|_{\mathcal{H}^m,\gamma} \leq C\,\|\widetilde{f}\|_{\mathcal{H}^m,\gamma}\,.$$

Enlarging γ_0 and C if necessary, it is sufficient to prove that

$$
(5.9) \quad \begin{aligned} &\gamma\|\widetilde{u}\|_{\mathcal{H}^{m},\gamma} + \sqrt{\varepsilon\gamma}\|\nabla_x\widetilde{u}\|_{\mathcal{H}^{m},\gamma} + \sqrt{\varepsilon}\|\partial_t\widetilde{u}\|_{\mathcal{H}^{m},\gamma} + \varepsilon^{3/2}\|\nabla_x^2\widetilde{u}\|_{\mathcal{H}^{m},\gamma} \\ &\qquad \leq C\,\|\widetilde{f}\|_{\mathcal{H}^{m},\gamma} + \|\widetilde{u}\|_{\mathcal{H}^{m},\gamma} + \sqrt{\varepsilon}\|\nabla_x\widetilde{u}\|_{\mathcal{H}^{m},\gamma}\,. \end{aligned}
$$

We show that this estimate, and therefore Theorem 5.1, follow from the next result.

PROPOSITION 5.3. *For all* $(\underline{t},\underline{x}) \in \mathbb{R}\times\overline{\Omega}$, *there are a neighborhood* ω *and constants* C *and* γ_0 *such that the estimate* (5.9) *is satisfied for all* $\gamma \geq \gamma_0$, $\varepsilon \in]0,1]$ *and* $\underline{u} \in C_0^\infty(\omega\cap\overline{\Omega})$ *which vanishes* $\omega\cap(\mathbb{R}\times\partial\Omega)$.

Assuming this proposition, we cover $[-T_1,T_1]\times\Omega$ by a finite number of open sets where the conclusion of the proposition holds. Refining the covering gives C, γ_0, a finite covering $\bigcup\Omega_j$ of $\overline{\Omega}$ and $\delta > 0$ such that the estimate (5.9) holds for $\gamma \geq 0$, $\varepsilon \in]0,1]$ and u supported in $\omega_{j,k}\cap\overline{\Omega}$ vanishing on $\omega_{j,k}\cap(\mathbb{R}\times\partial\Omega)$, with $\omega_{j,k} = [(k-1)\delta,(k+1)\delta]\times\Omega_j$, for all j and all $k\in[-k_1,k_1]$, where $(k_1-1)\delta \geq T_1$.

Because the coefficients of $\mathcal{P}^\varepsilon$ are independent of time for $t \geq T_1$, the estimate on $\omega_{j,k}$ is implied by the estimate on ω_{j,k_1} for $k > k_1$ since the equation is invariant by the translation $t\mapsto t-(k-k_1)\delta$ from $\omega_{j,k}$ to ω_{j,k_1}. Similarly, the estimate on $\omega_{j,k}$ is implied by the estimate on $\omega_{j,-k_1}$ for $k < k_1$. Therefore the estimate holds for all j and all $k \in \mathbb{Z}$.

Choose a partition of unity $\sum\chi_j(x) = 1$ on $\overline{\Omega}$, with $\chi_j \in C_0^\infty(\Omega_j)$. Choose next $\theta \in C_0^\infty(]-\delta,\delta[)$ such that $\sum_k\theta_k(t) = 1$ on $\mathbb{R}$ where $\theta_k(t) = \theta(t-k\delta)$. For all $\widetilde{u}\in C_0^\infty(\mathbb{R}\times\overline{\Omega})$ which vanishes on $\mathbb{R}\times\partial\Omega$, we can apply the estimate (5.9) to $\widetilde{u}_{j,k} = \chi_j(x)\theta_k(t)\widetilde{u}$. We note that

$$
\begin{aligned} \widetilde{\mathcal{P}}^\varepsilon\widetilde{u}_{j,k} = &\chi_j\theta_k\widetilde{f} + (\partial_t\theta_k)\chi_j\widetilde{u} + \sum_{p=1}^{d} A_p^\varepsilon\theta_k(\partial_p\chi_j)\widetilde{u} \\ &-\varepsilon\sum_{p,q=1}^{d} B_{p,q}^\varepsilon\theta_k\Big((\partial_p\chi_j)\partial_q\widetilde{u} + (\partial_q\chi_j)\partial_p\widetilde{u} + (\partial_{p,q}^2\chi_j)\widetilde{u}\Big)\,. \end{aligned}
$$

Therefore, since $\theta_k^\sharp = \theta_{k-1}+\theta_k+\theta_{k+1} = 1$ on the support of θ_k, one has

$$
\|\widetilde{\mathcal{P}}^\varepsilon\widetilde{u}_{j,k}\|_{\mathcal{H}^{m},\gamma} \leq \|\chi_j\theta_k\widetilde{f}\|_{\mathcal{H}^{m},\gamma} + C\Big(\|\theta_k^\sharp\widetilde{u}\|_{\mathcal{H}^{m},\gamma} + \varepsilon\|\theta_k^\sharp\nabla_x u\|_{\mathcal{H}^{m},\gamma}\Big)\,.
$$

The left hand side of (5.9) for $\widetilde{u}$, is clearly less than or equal to the sum of the left hand sides for the $\widetilde{u}_{j,k}$. Moreover, since the supports of the θ_k do not overlap three by three, one has

$$
\sum_{j,k}\|\widetilde{\chi}_j\theta_k\widetilde{f}\|_{\mathcal{H}^{m},\gamma} \leq C\|\widetilde{f}\|_{\mathcal{H}^{m},\gamma}\,, \quad \|\theta_k^\sharp\widetilde{u}\|_{\mathcal{H}^{m},\gamma} \leq C\|\widetilde{u}\|_{\mathcal{H}^{m},\gamma}\,,
$$

where C is independent of γ, u and f. There is a similar estimate for the x derivatives. Therefore, adding the different estimates (5.9) for the $\widetilde{u}_{j,k}$ yields an estimate (5.9) for $\widetilde{u}$, where the only modification is that the constant C is increased.

PROPOSITION 5.4. *For all* $(\underline{t},\underline{x}) \in \mathbb{R}\times\partial\Omega$, *there are a neighborhood* ω *and constants* C *and* γ_0 *such that the estimate* (5.9) *is satisfied for all* $\gamma \geq \gamma_0$, $\varepsilon \in]0,1]$ *and* $\widetilde{u} \in C_0^\infty(\omega\cap\overline{\Omega})$ *which vanishes* $\omega\cap(\mathbb{R}\times\partial\Omega)$.

PROOF. We consider local coordinates near $(\underline{t}, \underline{x})$ and use the notations of section 4, where x denotes the normal coordinates and $y = (y_1, \ldots, y_{d-1})$ the tangential variables.

With notations as in Theorem 4.1, we remark that the weight function φ defined in (4.11) satisfies

$$\varphi^2 \geq c\big(\gamma + \varepsilon|\eta|^2 + \min(\varepsilon\tau^2, |\tau|)\big) .$$

Thus

$$\varphi^2 \geq c'\big(\gamma + \sqrt{\varepsilon\gamma}|\eta| + \varepsilon|\eta|^2 + \sqrt{\varepsilon}|\tau|\big) , \quad \varphi \geq c'\big(\sqrt{\gamma} + \sqrt{\varepsilon}|\eta|\big) .$$

Therefore, for u supported in ω satisfying $u = 0$ on $\{x = 0\}$, since $v = \varepsilon\partial_x u$, Theorem 4.1 implies that

$$\begin{aligned}\gamma \|u\|_{\mathcal{H}^m,\gamma} + \sqrt{\varepsilon\gamma}\, \|\nabla_y u\|_{\mathcal{H}^m,\gamma} + \sqrt{\varepsilon}\, \|\partial_t u\|_{\mathcal{H}^m,\gamma} + \varepsilon\, \|\nabla_y^2 u\|_{\mathcal{H}^m,\gamma} \\ + \sqrt{\varepsilon\gamma}\, \|\partial_x u\|_{\mathcal{H}^m,\gamma} + \varepsilon\, \|\nabla_y\partial_x u\|_{\mathcal{H}^m,\gamma} \leq C\, \|f\|_{\mathcal{H}^m,\gamma} ,\end{aligned} \tag{5.10}$$

where, for simplicity, we have dropped the tildes. From the equation, we have

$$\begin{aligned}\varepsilon\partial_x^2 u = -\Phi_1^\varepsilon f + \Phi_2^\varepsilon(\partial_t u + \gamma u) + \Phi_3^\varepsilon \nabla_y u \\ + \varepsilon\Phi_4^\varepsilon \nabla_y^2 u + \Phi_5^\varepsilon \partial_x u + \varepsilon\Phi_6 \nabla_y \partial_x u + \frac{1}{\varepsilon} E^\varepsilon u ,\end{aligned} \tag{5.11}$$

where the Φ_j^ε are coefficients depending on u_a as in (4.7). In particular, by 4.1 they satisfy

$$\sup_\varepsilon \sup_{|\alpha| \leq m} \|Z^\alpha \Phi_j^\varepsilon\|_{L^\infty} < +\infty . \tag{5.12}$$

Multiply the identity (5.11) by $\varepsilon^{1/2}$. Then, all the terms in the right hand side are controlled by $\|f\|_{\mathcal{H}^{m,},\gamma}$ and the left hand side of (5.10), except

$$\varepsilon^{-1/2} E^\varepsilon u .$$

Here we use that $E^\varepsilon = E^\sharp + \varepsilon E_1^\varepsilon$ (see (4.4)) and $E^\sharp = e^{-\theta x/\varepsilon} E_2^\varepsilon$ where E_1^ε and E_2^ε satisfy (5.12). By (4.69), we have

$$\varepsilon^{-1/2} \|e^{-\theta x/\varepsilon} u\|_{\mathcal{H}^m,\gamma} \leq C\, \varepsilon^{1/2} \|\partial_x u\|_{\mathcal{H}^m,\gamma}$$

and thus

$$\varepsilon^{-1/2} \|E^\varepsilon u\|_{\mathcal{H}^m,\gamma} \leq C\, \big(\varepsilon^{1/2} \|\partial_x u\|_{\mathcal{H}^m,\gamma} + \sqrt{\varepsilon} \|u\|_{\mathcal{H}^m,\gamma}\big) .$$

With (5.11), we see that $\varepsilon^{3/2} \|\partial_x^2 u\|_{\mathcal{H}^m,\gamma}$ can be added to the left hand side, increasing the constant C. In order to have a formulation invariant under the change of coordinates, we give a symmetric role to the second normal and tangential derivatives, and (5.10) implies

$$\begin{aligned}\gamma\|u\|_{\mathcal{H}^m,\gamma} + \sqrt{\varepsilon\gamma}\|\nabla_{y,x} u\|_{\mathcal{H}^m,\gamma} + \sqrt{\varepsilon}\|\partial_t u\|_{\mathcal{H}^m,\gamma} \\ + \varepsilon^{3/2}\, \|\nabla_{y,x}^2 u\|_{\mathcal{H}^m,\gamma} \leq C\, \|f\|_{\mathcal{H}^m,\gamma} .\end{aligned}$$

Going back to the original coordinates, this implies (5.9) □

The next result implies that the statement in Proposition 5.3 is satisfied when $\underline{x} \in \Omega$, finishing the proof of this proposition and therefore of Theorem 5.1. The estimate we prove below is indeed a slight improvement of (5.9) for functions supported away from $\partial\Omega$.

PROPOSITION 5.5. *Suppose that Ω_1 is an open set such that $\overline{\Omega}_1 \subset \Omega$. Then, there are C and γ_0 such that for $\varepsilon \in]0,1]$, all $\gamma \geq \gamma_0$ and all u in $C_0^\infty(\mathbb{R} \times \Omega_1)$,*

$$\begin{aligned}\gamma\|u\|_{\mathcal{H}^m,\gamma} + \sqrt{\varepsilon\gamma}\|\nabla_x u\|_{\mathcal{H}^m,\gamma} + \sqrt{\varepsilon}\,\|\partial_t u\|_{\mathcal{H}^m,\gamma} + \varepsilon\|\nabla_x^2 u\|_{\mathcal{H}^m,\gamma} \\ \leq C\Big(\|(\mathcal{P}^\varepsilon + \gamma)u\|_{\mathcal{H}^m,\gamma} + \|u\|_{\mathcal{H}^m,\gamma} + \varepsilon\|\nabla_x u\|_{\mathcal{H}^m,\gamma}\Big).\end{aligned} \tag{5.13}$$

Introduce

$$\mathcal{P}^\flat = \partial_t + \gamma + \sum_{j=1}^d \widetilde{A}_j \partial_j - \varepsilon \sum_{j,k=1}^d \widetilde{B}_{j,k}\partial_j\partial_k\,, \tag{5.14}$$

where $\widetilde{A}$ stands for the function A evaluated at $(t, x, u_a^\varepsilon(t,x))$. For simplicity, we do not mention in the notation that this operator depends on the parameters ε and γ. The definition of the coefficients given after (4.3) shows that

$$\mathcal{P}^\varepsilon - \mathcal{P}^\flat = \sum_j A'_j \varepsilon \partial_j + E'\,,$$

where the coefficients have the form

$$A'_j = \widetilde{\Phi}\partial_k u_a\,, \quad E' = \sum \widetilde{\Phi}\partial_j u_a\,, + \sum \varepsilon\widetilde{\Phi}\partial_{j,k}^2 u_a + \sum \varepsilon\widetilde{\Phi}\partial_j u_a \partial_k u_a\,,$$

where the Φ's are smooth functions of (t, x, u) and $\widetilde{\Phi}$ stands for the evaluation of Φ at $u = u_a(t,x)$. Since $\overline{\Omega}_1$ is compact in Ω, the assumptions (4.1) on u_0 and v^ε imply that

$$\sup_\varepsilon \sup_{|\alpha|\leq m} \Big(\sum_j \|\partial_{t,x}^\alpha A'_j\|_{L^\infty(\mathbb{R}\times\Omega_1)} + \|\partial_{t,x}^\alpha E'\|_{L^\infty(\mathbb{R}\times\Omega_1)}\Big) < \infty. \tag{5.15}$$

Therefore,

$$\|(\mathcal{P}^\varepsilon - \mathcal{P}^\flat)u\|_{\mathcal{H}^m,\gamma} \lesssim \|u\|_{\mathcal{H}^m,\gamma} + \varepsilon\|\nabla_x u\|_{\mathcal{H}^m,\gamma}\,.$$

Moreover, the conormal vector fields generate all the derivatives on ω, and the spaces $\mathcal{H}^m(\omega)$ are the usual Sobolev spaces $H^m(\omega)$. Thus, introducing the norms

$$\|u\|_{m,\gamma} = \sum_{|\alpha|\leq m} \gamma^{m-|\alpha|}\|\partial_{t,x}^\alpha u\|_{L^2}\,, \tag{5.16}$$

we see that Proposition 5.5 follows from the estimate

$$\begin{aligned}\gamma\|u\|_{m,\gamma} + \sqrt{\varepsilon\gamma}\|\nabla_x u\|_{m,\gamma} + \sqrt{\varepsilon}\,\|\partial_t u\|_{m,\gamma} + \varepsilon\|\nabla_x^2 u\|_{m,\gamma} \\ \leq C\Big(\|\mathcal{P}^\flat u\|_{m,\gamma} + \|u\|_{m,\gamma} + \varepsilon\|\nabla_x u\|_{m,\gamma}\Big)\end{aligned} \tag{5.17}$$

for $u \in C_0^\infty(\mathbb{R} \times \Omega_1)$. We first prove this estimate for $m = 0$ and next prove it for general m, differentiating the equation.

5.3. Proof of Proposition 5.5. The analysis is quite similar to the analysis made in section 4, but much simpler since there are no boundary conditions, no glancing modes and no singular terms. However, the proof relies again on the construction of a (para-differential) symmetrizer, where we now use the "usual" Bony-Meyer paradifferential calculus (see Appendix B). To avoid repetitions, we just give now the general scheme of the analysis, leaving the details to the reader, who can easily fill in the gaps repeating the computations of section 4 and using the results recalled in Appendix B.

We start with a symbolic analysis. With obvious notations we write

$$\mathcal{P}^\flat = (\partial_t + \gamma)\mathrm{Id} + A^\flat(t,x,\partial_x) + \varepsilon B^\flat(t,x,\partial_x)\,. \tag{5.18}$$

The symbol of $A^\flat$ and $B^\flat$ are

$$\mathrm{a}^\varepsilon(t,x,\xi) = \sum_{j=1}^d i\xi_j\,\widetilde{A}_j(t,x)\,, \quad \mathrm{b}^\varepsilon(t,x,\xi) = \sum_{j=1,k}^d \xi_j\xi_k\,B_{j,k}(t,x)\,. \tag{5.19}$$

We look for symmetrizers $\mathrm{s}^\varepsilon(t,x,\xi)$ that are $N\times N$ matrix valued symbols of degree zero, i.e. such that

$$\begin{aligned} &\forall\alpha\in\mathbb{N}^d, \quad \exists C_\alpha, \quad \forall\varepsilon\in]0,1], \quad \forall(t,x,\xi)\in\mathbb{R}^{1+d}\times\mathbb{R}^d :\\ &\big|\partial_\xi^\alpha \mathrm{s}(t,x,\xi)\big| + \big|\partial_\xi^\alpha\nabla_{t,x}\mathrm{s}(t,x,\xi)\big| \le C(1+|\xi|)^{-|\alpha|}\,. \end{aligned} \tag{5.20}$$

In the sequel, we fix $\kappa\in C_0^\infty(\Omega)$, non negative and equal to one on $\overline{\Omega}_1$, and $\psi\in C^\infty(\mathbb{R}^d)$ equal to one for $|\xi|\ge 2$ and vanishing for $|\xi|\le 1$.

PROPOSITION 5.6. *There are families of symbols* $\mathrm{s}^\varepsilon(t,x,\xi)$ *and* r^ε *which satisfies* (5.20) *and*

$$\mathrm{s}^\varepsilon = (\mathrm{s}^\varepsilon)^* \quad \textit{and} \quad \kappa^2\mathrm{s}^\varepsilon \ge \kappa^2\mathrm{Id}\,, \tag{5.21}$$

$$\mathrm{Re}\,\big(\mathrm{s}^\varepsilon(\mathrm{a}^\varepsilon + \varepsilon\mathrm{b}^\varepsilon)\big) = \varepsilon|\xi|^2\mathrm{r}^\varepsilon\,, \quad \textit{with} \quad \kappa^2\mathrm{r}^\varepsilon \ge \kappa^2\,\psi^2\,\mathrm{Id}\,. \tag{5.22}$$

PROOF. The symbols a^ε and b^ε are the evaluation at $u = u_a^\varepsilon(t,x)$ and $b = b(t,x) = (t,x,b_0(t,x))$ of smooth symbols

$$\mathcal{A}(b,u,\xi) = \sum_{j=1}^d i\xi_j\,A_j(b,u)\,, \quad \mathcal{B}(b,u,\xi) = \sum_{j=1,k}^d \xi_j\xi_k\,B_{j,k}(b,u)\,.$$

We construct symmetrizers $\mathcal{S}^\varepsilon(b,u,\xi)$ and next choose $\mathrm{s}^\varepsilon = \kappa_1(t,x)\mathcal{S}(u_a^\varepsilon(t,x),\xi)$ with $\kappa_1\in C_0^\infty(\Omega)$, such that $\kappa\kappa_1 = \kappa$.

We proceed in two steps. When $\varepsilon|\xi|$ is small, we consider $\varepsilon\mathcal{B}$ as a perturbation of $\mathcal{A}$. Introduce the notations $\xi = |\xi|\hat{\xi}$, $\rho = \varepsilon|\xi|$ so that, forgetting the dependence on the parameter (b,u),

$$\mathcal{A}(\xi) + \varepsilon\mathcal{B}(\xi) = |\xi|\big(\mathcal{A}(\hat{\xi}) + \rho\mathcal{B}(\hat{\xi})\big)\,.$$

The constant multiplicity Assumption (H2) implies that there is a smooth invertible matrix $\mathcal{V}(\xi)$, homogeneous of degree zero, such that $\mathcal{V}^{-1}\mathcal{A}\mathcal{V}$ is block diagonal. Thus, there is $\rho_0>0$ such that for $\rho\in[0,\rho_0[$, there is $\mathcal{V}(\xi,\rho)$ such that

$$\mathcal{V}^{-1}(\hat{\xi},\rho)\big(\mathcal{A}(\hat{\xi}) + \rho\mathcal{B}(\hat{\xi})\big)\mathcal{V}(\hat{\xi},\rho) = \begin{pmatrix} i\lambda_1\mathrm{Id}+\rho B_1 & 0 & \\ 0 & \ddots & 0\\ & 0 & i\lambda_\nu\mathrm{Id}+\rho B_\nu \end{pmatrix}.$$

Moreover, Assumption (H3) implies that the eigenvalues of the blocks $B_k(\hat{\xi},\rho)$ remain in $\mathrm{Re}\,\mu\ge c$ when $\hat{\xi}$ is in the unit sphere and ρ remains small. Therefore, there are matrices $S_j(\hat{\xi},\rho)$ such that $S_j = S_j^*$ and $\mathrm{Re}\,S_jB_j$ are uniformly positive definite. With

$$\widehat{\mathcal{S}}(\hat{\xi},\rho) = \begin{pmatrix} S_1 & 0 & \\ 0 & \ddots & 0\\ & 0 & S_\nu \end{pmatrix}$$

and

$$\mathcal{S}_1^\varepsilon(\xi) = \psi(\xi)\widehat{\mathcal{S}}(\frac{\xi}{|\xi|}, \varepsilon|\xi|)$$

we obtain a symmetrizer for $\varepsilon|\xi| \le \rho_0$.

When $\varepsilon|\xi| \ge \rho_0/2$, we consider that the leading term is $\varepsilon\mathcal{B}$ and write

$$\mathcal{A}(\xi) + \varepsilon\mathcal{B}(\xi) = \varepsilon|\xi|^2\big(\lambda\mathcal{A}(\hat{\xi}) + \mathcal{B}(\hat{\xi})\big)$$

with $\lambda = 1/\varepsilon|\xi| \le \lambda_0 := 2/\rho_0$. By Assumption (H3), the eigenvalues of $\lambda\mathcal{A}(\hat{\xi})+\mathcal{B}(\hat{\xi})$ remain in $\operatorname{Re}\mu \ge c > 0$ and thus there is a symmetrizer $\widetilde{\mathcal{S}}(\hat{\xi},\lambda)$ for $\lambda\mathcal{A}(\hat{\xi}) + \mathcal{B}(\hat{\xi})$. With

$$\mathcal{S}_2^\varepsilon(\xi) = \widetilde{\mathcal{S}}(\frac{\xi}{|\xi|}, 1/\varepsilon|\xi|)$$

we obtain a symmetrizer for $\varepsilon|\xi| \ge \rho_2$.

We paste the two symmetrizers, defining

$$\mathcal{S}^\varepsilon(\xi) = \chi(\xi/\rho_0)\mathcal{S}_1^\varepsilon(\xi) + (1 - \chi(\xi/\rho_0)\,\mathcal{S}_2^\varepsilon(\xi)\,,$$

with $\chi \in C_0^\infty(\{|\xi| \le 1\}$ equal to one for $|\xi| \le 1/2$.

The construction holds as long as the parameters (b, u) remain in a compact of the set $\mathcal{O}$ where Assumptions (H2) and (H3) are satisfied. Thus, it holds for $b = b(t,x)$ and $u = u_{\mathrm{a}}^\varepsilon(t,x)$ when x remains in the support of κ and ε is small enough (note that $u_{\mathrm{a}}^\varepsilon$ is independent of time when t is large).

Thus $\mathrm{s}^\varepsilon(t,x,\xi) = \mathcal{S}^\varepsilon(u_{\mathrm{a}}^\varepsilon(t,x),\xi)$ satisfies the properties listed in Proposition 5.6. When $\varepsilon \ge \varepsilon_0 > 0$, it is sufficient to use Assumption (H1), because in this case $\mathcal{A}$ is a bounded perturbation of $\varepsilon\mathcal{B}$ when $|\xi| \ge 1$.

Since $\kappa_1 = 0$ near $\partial\Omega$, note that $u_{\mathrm{a}}^\varepsilon$ is smooth in the support of κ_1 so that x derivatives are allowed as in estimate (5.20). □

PROPOSITION 5.7. *There are constants C and γ_0 such that the estimate* (5.17) *with $m = 0$ holds all $\varepsilon \in]0,1]$, all $\gamma \ge \gamma_0$ and $u \in C_0^\infty(\mathbb{R}\times\Omega_1)$.*

PROOF. We use para-differential operators acting on functions of in $x \in \mathbb{R}^d$, t being considered as a parameter (see Appendix B). For a symbol a, we denote by T_a the corresponding operator.

We fix a cut off function $\kappa_1 \in C_0^\infty(\Omega)$, which is equal to one on the support of κ, and thus on Ω_1. Therefore, for $u \in C_0^\infty(\mathbb{R}\times\Omega_1)$, $\mathcal{P}^b u = \kappa_1\mathcal{P}^b u$. The coefficients $\kappa_1\widetilde{A}_j$ and $\kappa_1\widetilde{B}_{j,k}$, extended by zero outside Ω, are bounded in $W^{1,\infty}(\mathbb{R}^{1+d})$. Therefore

$$\|A^b u - T_{\kappa_1 \mathrm{a}^\varepsilon} u\|_0 \lesssim \|u\|_0\,, \quad \|B^b u - T_{\kappa_1 \mathrm{b}^\varepsilon} u\|_0 \lesssim \|\nabla_x u\|_0$$

and it is sufficient to prove (5.17) with $\mathcal{P}^b u$ replaced by

$$f := (\partial_t + \gamma)u + T_{\kappa_1 \mathrm{a}^\varepsilon} u + \varepsilon T_{\kappa_1 \mathrm{b}^\varepsilon} u\,. \tag{5.23}$$

We use the symmetrizer $S = \gamma\Sigma - \varepsilon\sum\partial_{x_j}\Sigma\partial_{x_j}$, where $\Sigma = \operatorname{Re} T_{\mathrm{s}^\varepsilon}$ and s^ε is the family of symbols given by Proposition 5.6. S is self adjoint in $L^2(\mathbb{R}^{1+d})$. Moreover, $[\partial_t, \Sigma] = \operatorname{Re} T_{\partial_t \mathrm{s}^\varepsilon}$ and $\partial_t \mathrm{s}^\varepsilon$ is a bounded family of symbols of degree zero. Thus

$$\operatorname{Re}\big(\!\big(S(\partial_t + \gamma)u\,,\, u\big)\!\big) = \gamma^2\big(\!\big(\Sigma u, u\big)\!\big) + \varepsilon\gamma\sum_{j=1}^d \big(\!\big(\Sigma\partial_j u, \partial_j u\big)\!\big) + err_1\,,$$

where $err_1 = \big(\!\!\big(([S, \partial_t]u, u \big)\!\!\big)$ satisfies

$$|err_1| \lesssim \gamma \|u\|_0^2 + \varepsilon \|\nabla_x u\|_0^2 . \tag{5.24}$$

The symbolic calculus (see Appendix B) implies that $\Sigma T_{\kappa_1 \mathrm{a}^\varepsilon} = T_{\kappa_1 \mathrm{s}^\varepsilon \mathrm{a}^\varepsilon}$ is of degree zero while $[\partial_j, T^\varepsilon_{\kappa_1 \mathrm{a}^\varepsilon}] = T_{\partial_j(\kappa_1 \mathrm{a}^\varepsilon)}$ is of degree one. Thus,

$$\mathrm{Re}\, \big(\!\!\big(ST_{\kappa_1 \mathrm{a}^\varepsilon} u, u) \big)\!\!\big) = \gamma \mathrm{Re}\, \big(\!\!\big(T_{\kappa_1 \mathrm{s}^\varepsilon \mathrm{a}^\varepsilon} u, u \big)\!\!\big) + \varepsilon \sum_{j=1}^d \big(\!\!\big(T_{\kappa_1 \mathrm{s}^\varepsilon \mathrm{a}^\varepsilon} \partial_j u, \partial_j u \big)\!\!\big) + err_2 ,$$

where err_2 satisfies (5.24). Similarly, since b^ε is of degree two,

$$\mathrm{Re}\, \big(\!\!\big(ST_{\kappa_1 \mathrm{b}^\varepsilon} u, u) \big)\!\!\big) = \gamma \mathrm{Re}\, \big(\!\!\big(T_{\kappa_1 \mathrm{s}^\varepsilon \mathrm{b}^\varepsilon} u, u \big)\!\!\big) + \varepsilon \sum_{j=1}^d \big(\!\!\big(T_{\kappa_1 \mathrm{s}^\varepsilon \mathrm{b}^\varepsilon} \partial_j u, \partial_j u \big)\!\!\big) + err_3 ,$$

where

$$|err_3| \lesssim \gamma \|u\|_0 \|\nabla_x u\|_0 + \varepsilon \|\nabla_x u\|_0 \|\nabla_x^2 u\|_0 .$$

Next we use that $\mathrm{s}^\varepsilon(\mathrm{a}^\varepsilon + \varepsilon \mathrm{b}^\varepsilon) = \varepsilon \mathrm{r}^\varepsilon |\xi|^2$ and that $T_{\kappa_1 \mathrm{r}^\varepsilon |\xi|^2} + \sum_k \partial_k T_{\kappa_1 \mathrm{r}^\varepsilon} \partial_k$ is of order one. Thus, adding the various estimates above, we get that

$$\gamma^2 \big(\!\!\big(\Sigma u, u \big)\!\!\big) + \varepsilon\gamma \sum_{j=1}^d \big(\!\!\big(\Sigma \partial_j u, \partial_j u \big)\!\!\big) + \gamma\varepsilon \sum_{k=1}^d \mathrm{Re}\, \big(\!\!\big(T_{\kappa_1 \mathrm{r}^\varepsilon} \partial_k u, \partial_k u \big)\!\!\big)$$
$$+ \varepsilon^2 \sum_{j,k=1}^d \big(\!\!\big(T_{\kappa_1 \mathrm{r}^\varepsilon} \partial_j \partial_k u, \partial_j \partial_k u \big)\!\!\big) = \mathrm{Re}\, \big(\!\!\big(Sf, u \big)\!\!\big) + err$$

with

$$|err| \lesssim \big(\gamma \|u\|_0 + \varepsilon \|\nabla_x^2 u\|_0 \big) \big(\|u\|_0 + \varepsilon \|\nabla_x u\|_0 \big)$$

(we have used that $\|\nabla_x u\|_0^2 \leq \|u\|_0 \|\nabla_x^2 u\|_0$). Moreover,

$$|\big(\!\!\big(Sf, u \big)\!\!\big)| = |\big(\!\!\big(f, Su \big)\!\!\big)| \lesssim \|f\|_0 \big(\gamma \|u\|_0 + \varepsilon \|\nabla_x^2 u\|_0 \big) .$$

For $v \in C_0^\infty(\mathbb{R} \times \Omega_1)$, writing that $v = \kappa v$, the positivity conditions (5.21) and (5.22) imply that

$$\gamma \|v\|_0^2 \lesssim \big(\!\!\big(\Sigma v, v \big)\!\!\big) + \|v\|_0^2 , \quad \|\partial_j v\|_0^2 \lesssim \big(\!\!\big(\Sigma \partial_j v, \partial_j v \big)\!\!\big) + \|v\|_0^2$$

and

$$\|\partial_k v\|_0^2 \lesssim \big(\!\!\big(T_{\kappa_1 \mathrm{s}^\varepsilon} \partial_k v, \partial_k v \big)\!\!\big) + \|v\|_0^2 .$$

Therefore, we have proved that

$$\gamma^2 \|u\|_0^2 + \varepsilon\gamma \|\nabla_x u\|_0^2 + \varepsilon^2 \|\nabla_x^2\|_0 \lesssim \big(\|f\|_0 + \|u\|_0 + \|\nabla_x u\|_0 \big) \big(\gamma \|u\|_0 + \varepsilon \|\nabla_x^2 u\|_0 \big)$$

and thus

$$\gamma \|u\|_0 + \sqrt{\varepsilon\gamma} \|\nabla_x u\|_0 + \varepsilon \|\nabla_x^2\| \lesssim \|f\|_0 + \|u\|_0 + \|\nabla_x u\|_0 . \tag{5.25}$$

Moreover, (5.23) implies that

$$\|\partial_t u\|_0 \lesssim \|f\|_0 + \gamma \|u\|_0 + \|\nabla_x u\|_0 + \varepsilon \|\nabla_x^2 u\|_0$$

and (5.25) also provides an estimate for $\sqrt{\varepsilon} \|\partial_t u\|_0$. Adding up, we obtain the estimate (5.17) for $m = 0$. □

PROOF OF PROPOSITION 5.5. It remains to prove that the estimate (5.17) holds for all $m \geq 1$. We differentiate the equation in space time. Using that the derivatives of the coefficients of $A^\flat$ and $B^\flat$ up to order m are bounded on $\mathbb{R} \times \Omega_1$, one has for all $|\alpha| \leq m$, $\mathcal{P}^\flat \partial_{t,x}^\alpha u = \partial_{t,x}^\alpha \mathcal{P}^\flat u + g_\alpha$, with

$$\gamma^{m-|\alpha|} \|g_\alpha\|_0 \lesssim \|u\|_{m,\gamma} + \varepsilon \|\nabla_x u\|_{m,\gamma} .$$

Applying Proposition 5.7 to the derivatives $\partial_{t,x}^\alpha u$ and adding the various estimates one obtain (5.17). □

5.4. Proof of the L^∞ estimates. We now prove the estimates (1.32), finishing the proof of Theorem 1.10.

THEOREM 5.8. *If $m > \frac{d+1}{2} + 2$, there is a constant C such that for all $\varepsilon \in]0,1]$, all $f \in \mathcal{H}^m([-T_0, T_0] \times \Omega)$ vanishing for $t < 0$, the solution u of (5.1) which vanishes for $t < 0$, satisfies*

$$\sum_{|I| \leq 2} \|Z_I u\|_{L^\infty} + \varepsilon \sum_{|I| \leq 1} \|Z_I \nabla_{t,x} u\|_{L^\infty} \leq C \|f\|_{\mathcal{H}^m} . \tag{5.26}$$

If in addition $f \in L^\infty([-T_0, T_0] \times \Omega)$, then

$$\varepsilon^2 \|\nabla_x^2 u\|_{L^\infty} \leq C\big(\|f\|_{\mathcal{H}^m} + \varepsilon \|f\|_{L^\infty}\big) . \tag{5.27}$$

PROOF. **a)** By Theorem 1.10, we already know that

$$\|u\|_{\mathcal{H}^m} + \sqrt{\varepsilon}\, \|\nabla_x u\|_{\mathcal{H}^m} \lesssim \|f\|_{\mathcal{H}^m} . \tag{5.28}$$

Because $m > \frac{d+1}{2} + 2$, the Sobolev embedding implies that for all open set Ω_1 with $\overline{\Omega}_1 \subset \Omega$, one has

$$\|u\|_{W^{2,\infty}([-T_0,T_0] \times \Omega_1)} \leq C \|u\|_{\mathcal{H}^m} .$$

Therefore, it is sufficient to prove the L^∞ estimates near the boundary.

Near $\underline{y} \in \partial\Omega$, consider a coordinate patch with coordinates $(x, y) \in \mathbb{R} \times \mathbb{R}^{d-1}$ such that Ω is defined by $\{x > 0\}$. With $\kappa \in C_0^\infty(\omega)$, $u_1 = \kappa u$ satisfies $\mathcal{P}^\varepsilon u_1 = f$ with $f_1 = \kappa f + [cP^\varepsilon, \kappa] u \in \mathcal{H}^m$ and, using (5.28), $\|f_1\|_{\mathcal{H}^m} \lesssim \|f\|_{\mathcal{H}^m}$. Therefore, to prove (5.26), it is sufficient to prove it for u_1, that is:

$$\sum_{|\alpha| \leq 2} \|Z^\alpha u_1\|_{L^\infty} + \varepsilon \sum_{|\alpha| \leq 1} \|Z^\alpha \nabla_{t,y,x} u_1\|_{L^\infty} \lesssim \|f\|_{\mathcal{H}^{m-1}} . \tag{5.29}$$

b) In the local coordinates, the equation reads

$$-\varepsilon \partial_x^2 u_1 + A^\varepsilon \partial_x u_1 + \frac{1}{\varepsilon} M^\varepsilon u_1 = (B_{d,d}^\varepsilon)^{-1} f_1 \tag{5.30}$$

(see (4.7)). We put in the right hand side all the derivatives $\partial_t u_1$ $\partial_y u_1$, $\varepsilon \partial_x \partial_y u_1$ and $\varepsilon \partial_y^2 u_1$, which satisfy

$$\|\partial_{t,y} u\|_{\mathcal{H}^{m-1}} + \varepsilon \|\partial_y \partial_{x,y} u\|_{\mathcal{H}^{m-1}} \lesssim \|u\|_{\mathcal{H}^m} + \varepsilon \|\nabla_{y,x} u\|_{\mathcal{H}^m} \leq \|f\|_{\mathcal{H}^m} .$$

Hence, using the notations introduced in (2.6), we obtain

$$-\varepsilon \partial_x^2 u_1 + A_0^\varepsilon \partial_x u_1 + \frac{1}{\varepsilon} E_0^\varepsilon u_1 = \widetilde{f}_1 \tag{5.31}$$

with

$$\|\widetilde{f}_1\|_{\mathcal{H}^{m-1}} \lesssim \|f\|_{\mathcal{H}^m}$$

and

$$A_0^\varepsilon = (B_{d,d}^\varepsilon)^{-1} A_d^\sharp, \quad E_0^\varepsilon = (B_{d,d}^\varepsilon)^{-1} E^\sharp .$$

We write this equation as a first order system:

$$\partial_x U_1 = \frac{1}{\varepsilon} G_0^\varepsilon U_1 + F_1 \tag{5.32}$$

with

$$U_1 = \begin{pmatrix} u_1 \\ v_1 \end{pmatrix}, \quad G_0^\varepsilon = \begin{pmatrix} 0 & \mathrm{Id} \\ E_0^\varepsilon & A_0^\varepsilon \end{pmatrix}, \qquad F_1 = \begin{pmatrix} 0 \\ -\widetilde{f}_1 \end{pmatrix}.$$

We prove that

$$\sum_{|\alpha|\le 2} \|Z^\alpha U_1\|_{L^\infty} \lesssim \|F\|_{\mathcal{H}^{m-1}} + \|u_1\|_{\mathcal{H}^m} + \frac{1}{\sqrt{\varepsilon}}\|v_1\|_{\mathcal{H}^m}. \tag{5.33}$$

By (5.28), the right hand side is $\lesssim \|f\|_{\mathcal{H}^m}$ and therefore (5.33) implies (5.29).

c) With notations as in sections 2 and 4, one has

$$G_0^\varepsilon(t,x) = \mathcal{G}(\frac{x}{\varepsilon}, p^\varepsilon(t,y,x), 0).$$

Moreover, $\mathcal{G}_0(z,p) := \mathcal{G}(z,p,0)$ converges at an exponential rate to a limit $\mathcal{G}_0^\infty(p)$ as z tends to infinity and the limit has the form

$$\mathcal{G}_0^\infty(p) = \begin{pmatrix} 0 & \mathrm{Id} \\ 0 & \mathcal{D}(p) \end{pmatrix}.$$

By Lemma 2.6, there is a smooth matrix $\mathcal{W}_0(z,p) = \mathcal{W}(z,p,0)$ such that $\mathcal{W}(z,p) - \mathrm{Id} = O(e^{-\theta z})$ and

$$\partial_z \mathcal{W}_0 = \mathcal{G}_0 \mathcal{W}_0 - \mathcal{W}_0 \mathcal{G}_0^\infty. \tag{5.34}$$

Moreover, by Lemma 2.9 there is a matrix $\mathcal{V}_0 = \mathcal{V}(p,0)$ such that

$$\mathcal{V}_0^{-1} \mathcal{G}_0^\infty \mathcal{V}_0 = \begin{pmatrix} 0 & 0 \\ 0 & \mathcal{D}(p) \end{pmatrix}, \quad \mathcal{V}_0 = \begin{pmatrix} \mathrm{Id} & \mathcal{D}(p) \\ 0 & \mathrm{Id} \end{pmatrix}. \tag{5.35}$$

With these notations, we introduce $\mathcal{R}_0(z,p) = \mathcal{W}_0(z,p)\mathcal{V}_0(p)$ and

$$R_0^\varepsilon(t,y,x) = \mathcal{R}_0(\frac{x}{\varepsilon}, p^\varepsilon(t,y,x)), \quad D_0^\varepsilon(t,y,x) = \mathcal{D}_0(p^\varepsilon(t,y,x)).$$

Introduce $U_2 = (R_0^\varepsilon)^{-1} U_1$. Then, (5.34) implies that

$$\partial_x U_2 = \frac{1}{\varepsilon} \begin{pmatrix} 0 & 0 \\ 0 & D^\varepsilon \end{pmatrix} U_2 + \begin{pmatrix} f_2 \\ g_2 \end{pmatrix} \tag{5.36}$$

where $F_2 := \begin{pmatrix} f_2 \\ g_2 \end{pmatrix} = F_1 + (\partial_x p^\varepsilon) \cdot \nabla_p \mathcal{R}^0(\frac{x}{\varepsilon}, p^\varepsilon) U_1$ satisfies

$$\|F_2\|_{\mathcal{H}^{m-1}} \lesssim \|F_1\|_{\mathcal{H}^{m-1}} + \|U_1\|_{\mathcal{H}^{m-1}}.$$

The commutators of R^ε and $(R^\varepsilon)^{-1}$with Z^α are bounded for $|\alpha| \le m$, thus

$$\sum_{|\alpha|\le 2} \|Z^\alpha U_1\|_{L^\infty} \lesssim \sum_{|\alpha|\le 2} \|Z^\alpha U_2\|_{L^\infty}, \quad \|U_2\|_{\mathcal{H}^m} \lesssim \|U_1\|_{\mathcal{H}^m}.$$

Moreover, (5.35) implies that $v_2 = v_1 + O^{-\theta x/\varepsilon} U_1$. Therefore,

$$\|v_2\|_{\mathcal{H}^m} \lesssim \|v_1\|_{\mathcal{H}^m} + \|e^{-\theta x/\varepsilon} u_1\|_{\mathcal{H}^m}.$$

Using that $u_{1|x=0} = 0$, (4.69) implies that $\|e^{-\theta x/\varepsilon} u_1\|_{\mathcal{H}^m} \lesssim \|\varepsilon \partial_x u_1\|_{\mathcal{H}^m} = \|v_1\|_{\mathcal{H}^m}$. Thus

$$\|u_2\|_{\mathcal{H}^m} + \frac{1}{\sqrt{\varepsilon}}\|v_2\|_{\mathcal{H}^m} \lesssim \|u_1\|_{\mathcal{H}^m} + \frac{1}{\sqrt{\varepsilon}}\|v_1\|_{\mathcal{H}^m}.$$

Therefore, (5.33) follows from the estimate

$$\sum_{|\alpha|\le 2} \|Z^\alpha U_2\|_{L^\infty} \lesssim \|F_2\|_{\mathcal{H}^{m-1}} + \|u_2\|_{\mathcal{H}^m} + \frac{1}{\sqrt{\varepsilon}}\|v_2\|_{\mathcal{H}^m} . \tag{5.37}$$

d) Introduce the Sobolev spaces of tangentially smooth functions: we say that $u \in H^m_{tg}$ if the tangential derivatives $\partial^\alpha_{t,y} u$ of order $|\alpha| \le m$ belong to $L^2([-T_0, T_0] \times \mathbb{R}^d_+)$, equipped with the obvious norm. Then $\mathcal{H}^m \subset H^m_{tg}$ and

$$\|u\|_{H^m_{tg}} \le \|u\|_{\mathcal{H}^m} .$$

Next, we use the following lemma

LEMMA 5.9. *If $s > \frac{d+1}{2}$, there is C such that for all $u \in H^s_{tg}$ such that $\partial_x u \in H^{s-1}_{tg}$, one has $u \in L^\infty([-T_0, T_0] \times \mathbb{R}^d_+)$ and*

$$\|u\|^2_{L^\infty} \le C \|u\|_{H^s_{tg}} \|\partial_x u\|_{H^{s-1}_{tg}} .$$

Applied to $Z^\alpha u$ for $|\alpha| \le 2$, since $m - 2 > \frac{d+1}{2}$, this lemma implies that

$$\|Z^\alpha u\|^2_{L^\infty} \lesssim \|u\|_{\mathcal{H}^m} \|\partial_x u\|_{\mathcal{H}^{m-1}} . \tag{5.38}$$

By (5.36),

$$\|\partial_x u_2\|_{\mathcal{H}^{m-1}} = \|f_2\|_{\mathcal{H}^{m-1}} \quad \text{and} \quad \|\partial_x v_2\|_{\mathcal{H}^{m-1}} \le \frac{1}{\varepsilon}\|v_2\|_{\mathcal{H}^{m-1}} + \|g_2\|_{\mathcal{H}^{m-1}} .$$

Therefore, (5.38) implies that

$$\|Z^\alpha u_2\|^2_{L^\infty} \lesssim \|u\|^2_{\mathcal{H}^m} + \|f_2\|^2_{\mathcal{H}^{m-1}} , \quad \|Z^\alpha v_2\|^2_{L^\infty} \lesssim \frac{1}{\varepsilon}\|v_2\|^2_{\mathcal{H}^m} + \|g_2\|^2_{\mathcal{H}^{m-1}} .$$

The estimate (5.37) follows and the proof of (5.29) is complete.

e) It remains to prove (5.27) near the boundary, that is for localized functions u_1. The only missing estimate is for the second normal derivative $\varepsilon^2 \partial_x u_1$. Using (5.30), one obtains

$$\begin{aligned}\varepsilon^2 \|\partial^2_x u_1\|_{L^\infty} \lesssim \varepsilon \|f\|_{L^\infty} + \|u\|_{L^\infty} + \varepsilon \|\nabla_{t,y,x} u_1\|_{L^\infty} \\ + \varepsilon^2 \|\nabla^2_y u_1\|_{L^\infty} + \varepsilon^2 \|\nabla_y \partial_x u_1\|_{L^\infty} .\end{aligned}$$

With (5.29) this implies (5.28) and the proofs of Theorems 5.8 and are complete. □

6. Nonlinear stability

In this section we prove Theorem 1.11. Consider integers $m > \frac{d+1}{2}$ and $s_0 > m + 3 + \frac{d+1}{2}$. Consider the hyperbolic boundary value problem (1.1) (1.8):

$$L(b, u, \partial)u := \partial_t u + \sum_{j=1}^{d} A_j(b, u)\partial_j u = F(b, u) , \quad u_{|[-T_0,T_0]\times\partial\Omega} \in \mathcal{C} \tag{6.1}$$

with a forcing term $F(b, u)$ such that $F(0, 0) = 0$ and $b \in H^{s_0}([-T_0, T_0] \times \Omega)$ such that $b = 0$ for $t < 0$. We assume that the state $u = 0$ belongs the domain of hyperbolicity $\mathcal{O}$ in Assumption 1.1. The Assumption 1.4 implies that (6.1) satisfies the uniform Lopatinski condition. Shrinking T_0 if necessary, consider a solution $u_0 \in H^{s_0}([-T_0, T_0] \times \Omega)$ of the mixed Cauchy problem (1.1) (1.8) which vanishes for $t < 0$. Since $s > \frac{d+1}{2} + m + 3$, one has

$$u_0 \in W^{m+3,\infty}([-T_0, T_0] \times \Omega) , \qquad b_0 \in W^{m+3,\infty}([-T_0, T_0] \times \Omega). \tag{6.2}$$

Consider

$$u_0^\varepsilon(t,x) = W\big(b(t,x), u_0(t,x), \varphi(x)/\varepsilon\big)\,. \tag{6.3}$$

Then u_0^ε vanishes for $t<0$ and thus is an exact solution on $[-T_0, 0]\times\Omega$ of

$$L(b,u,\partial)u - \varepsilon\sum_{j,k=1}^{d}\partial_j\big(B_{j,k}(b,u)\partial_k u\big) = F(b,u)\,,\quad u_{|[-T_0,T_0]\times\partial\Omega} = 0\,. \tag{6.4}$$

We assume that for all $(t,x)\in[-T_0,T_0]\times\overline{\Omega})$, $(b(t,x), u_0(t,x))$ remains in a compact subset of $\mathcal{O}$ where the Assumptions 1.1, 1.2 and 1.4 are satisfied.

THEOREM 6.1. *There is $\varepsilon_0>0$ such that for all $\varepsilon\in]0,\varepsilon_0]$ the problem (6.4) has a unique solution u^ε which vanishes for $t<0$. Moreover,*

$$\|u^\varepsilon - u_0^\varepsilon\|_{\mathcal{H}^m} + \|u - u_0^\varepsilon\|_{L^\infty} = O(\varepsilon)\,. \tag{6.5}$$

We first construct a corrector u_1^ε such that $u_a^\varepsilon = u_0^\varepsilon + \varepsilon u_1^\varepsilon$ is a solution of (6.1) up to an error of size $O(\varepsilon)$.

LEMMA 6.2. *There is a family u_1^ε in $W^{m+2,\infty}([-T_0,T_0]\times\Omega)$ such that $u_0^\varepsilon = 0$ on $[-T_0,T_0]\times\partial\Omega$ and on $\{t<0\}$,*

$$\sup_\varepsilon\sup_{|J|\le m}\Big(\|Z_J u_1^\varepsilon\|_{L^\infty} + \varepsilon\|\nabla_{t,x}Z_J u_1^\varepsilon\|_{L^\infty} + \varepsilon^2\|\nabla_x^2 Z_J u_1^\varepsilon\|_{L^\infty}\Big) < \infty\,, \tag{6.6}$$

and $u_0^\varepsilon = u_0^\varepsilon + \varepsilon u_1^\varepsilon$ satisfies

$$L(b,u_a^\varepsilon,\partial)u_a^\varepsilon - \varepsilon\sum_{1\le j,k\le d}\partial_j\big(B_{j,k}(b,u_a^\varepsilon)\partial_k u_a^\varepsilon\big) - F(b,u_a^\varepsilon) = \varepsilon f^\varepsilon\,, \tag{6.7}$$

with

$$\sup_{\varepsilon\in]0,1]}\Big(\|f^\varepsilon\|_{\mathcal{H}^m} + \|f^\varepsilon\|_{L^\infty}\Big) < +\infty\,. \tag{6.8}$$

PROOF. By definition

$$u_0^\varepsilon - u_0 = W'(x,b,u_0,\varphi/\varepsilon)\,,$$

where $W'(x,u,z)$ is a smooth function of its arguments which converges at an exponential rate to zero when z tends to infinity. When u_0^ε is substituted in a smooth function $A(b,u)$, one has

$$A\big(b(t,x),u_0^\varepsilon(t,x)\big) = A\big(b(t,x),u_0(t,x)\big) + A'\big(b(t,x),u_0(t,x),\varphi(x)/\varepsilon\big)\,,$$

where A' is a smooth function of (x,b,u,z), exponentially decaying at infinity in z. This implies that

$$\begin{aligned}&L(b,u_0^\varepsilon,\partial)u_0^\varepsilon - \varepsilon\sum_{1\le j,k\le d}\partial_j\big(B_{j,k}(b,u^\varepsilon)\partial_k u^\varepsilon\big) - F(b,u^\varepsilon)\\ &\qquad = \frac{1}{\varepsilon}R_0(b,u_0,\frac{\varphi}{\varepsilon}) + R'(q(t,x),\frac{\varphi}{\varepsilon}) + \varepsilon f_0^\varepsilon\,,\end{aligned} \tag{6.9}$$

where

$$R_0(b,u_0,z) = A_n(b,W)\partial_z W - \partial_z\big(B_n(b,W)\partial_z W\big)\,.$$

$R'(q,z)$ is a smooth function of its arguments, exponentially decaying at infinity in z, and $q=(b,u_0,\partial_{t,x}b,\partial_{t,x}u)$. Moreover

$$f_0^\varepsilon = -\sum_{j,k=1}^{d}\partial_j\big(B_{j,k}(b,u_0)\partial_k u_0\big) + \widetilde{R}'(x,\widetilde{q},\varphi(x)/\varepsilon)\,,$$

where $\widetilde{R}'$ is similar to R' with now $\widetilde{q} = (b, u_0, \partial_{t,x} b, \partial_{t,x} u, \partial_x^2 b, \partial_x^2 u_0)$. In particular,

$$\sup_{\varepsilon \in]0,1]} \Big(\| f_0^\varepsilon \|_{\mathcal{H}^m} + \| f_0^\varepsilon \|_{L^\infty} \Big) \ < \ +\infty \, .$$

By Lemma 1.8, $R_0 = 0$ when $(b, u) \in \mathcal{C}$. Because u_0 satisfies the boundary condition, this function vanishes when $b = b(t, x)$, $u = u_0(t, x)$ and $x \in \partial\Omega$. Thus one can factor out x in $R_0(b(t, x), u_0(t, x), z)$ and, since W is exponentially decaying

$$\frac{1}{\varepsilon} R_0(b(t, x), u_0(t, x), \frac{\varphi}{\varepsilon}) = R_1'(q(t, x), \frac{\varphi}{\varepsilon}) \, .$$

Define $R_1 = R' + R_1'$.

We look for u_1^ε as a function

$$u_1^\varepsilon(t, x) = \mathcal{W}_1(q(t, x), \varphi(x)/\varepsilon) \tag{6.10}$$

with $\mathcal{W}_1$ C^∞ in the variables (q, z), and exponentially converging to a limit at $z = \infty$. With (6.9), we get that the left hand side of (6.7) is

$$(R_1 + \mathcal{L}\mathcal{W}_1)(x, q, \varphi/\varepsilon) \, + \, \varepsilon f_0^\varepsilon \, + \, \varepsilon R_1^\varepsilon(x, \widetilde{\widetilde{q}}, \varphi/\varepsilon) \, ,$$

where $\widetilde{\widetilde{q}} = (q, \partial_{t,x} q, \partial_x^2 q)$ belongs to $W^{m,\infty}$ and $\mathcal{L}$ is the linearized operator defined in (1.9):

$$\begin{aligned} \mathcal{L}\mathcal{W}_1 = A_n(\mathcal{W})\partial_z \mathcal{W}_1 + & (A_n'(\mathcal{W}) \cdot \mathcal{W}_1)\partial_z \mathcal{W} \\ & - \partial_z \Big(B_n(\mathcal{W})\partial_z \mathcal{W}_1 + (B_n'(\mathcal{W}) \cdot \mathcal{W}_1)\partial_z \mathcal{V} \Big) \, , \end{aligned}$$

where $A_n = \sum \partial_j \varphi A_j$, $B_n = \sum \partial_j \varphi \partial_j k B_{j,k}$ and the coefficients also depend on the parameters (x, b). Since R' is exponentially decaying at infinity, the Assumption 1.2 implies that the equation

$$\mathcal{L}\mathcal{W}_1 = -R' , \quad \mathcal{W}_{1|z=0} = 0 \, , \tag{6.11}$$

has solutions $\mathcal{W}_1$ which converge at an exponential rate at infinity. With this choice, the left hand side of (6.9) is $\varepsilon f^\varepsilon$ with $f^\varepsilon = f_0^\varepsilon + R^\varepsilon(x, \widetilde{\widetilde{q}}, \varphi/\varepsilon)$ which satisfies (6.8).

In addition, since $q \in W^{m+2,\infty}$ and $\mathcal{W}_1$ is smooth, the estimates (6.6) are satisfied. □

REMARK 6.3. In [**Gr-Gu**],the authors construct approximate solutions at all order, using BKW expansions. The construction of $\mathcal{W}_1$ is just one piece of their construction of the first corrector.

Next we solve the equation (6.4), looking for a solution $u^\varepsilon = u_a^\varepsilon + \varepsilon v^\varepsilon$. The equation for v^ε reads

$$\mathcal{P}_{u_a^\varepsilon} v^\varepsilon + \mathcal{Q}^\varepsilon(v^\varepsilon) \, = f^\varepsilon \, , \tag{6.12}$$

where $\mathcal{P}_{u_0^\varepsilon}$ is the linearized operator defined in (5.1) and $\mathcal{Q}^\varepsilon$ is a family of second order nonlinear operators acting on v^ε. An examination of the expansions, shows that $\mathcal{Q}^\varepsilon(v^\varepsilon)$ is a sum of terms of the form

$$\begin{aligned} \mathcal{Q}_1 &= \varepsilon \Phi(b, u_a^\varepsilon, \varepsilon v^\varepsilon) \, v^\varepsilon \, \partial_j v^\varepsilon \, , \\ \mathcal{Q}_2 &= \varepsilon \Phi(b, u_a^\varepsilon, \varepsilon v^\varepsilon) \, v^\varepsilon \, v^\varepsilon \, \partial_j u_a^\varepsilon \, , \\ \mathcal{Q}_3 &= \varepsilon^2 \partial_k \big(\Phi(b, u_a^\varepsilon, \varepsilon v^\varepsilon) \, v^\varepsilon \partial_j v^\varepsilon \big) \, , \\ \mathcal{Q}_4 &= \varepsilon^2 \partial_k \big(\Phi(b, u_a^\varepsilon, \varepsilon v^\varepsilon) \, v^\varepsilon \, v^\varepsilon \, \partial_j u_a^\varepsilon \big) \, , \end{aligned}$$

where the Φ's are smooth functions of their arguments, $\mathcal{Q}_1$ and $\mathcal{Q}_3$ stand for bilinear expressions in v^ε and $\partial_j v^\varepsilon$, while $\mathcal{Q}_2$ and $\mathcal{Q}_4$ are bilinear in v^ε and linear in $\partial_j u^\varepsilon_a$. Moreover, indices j and k run in $\{1, \dots, d\}$, which means that only spatial derivatives are present in $\mathcal{Q}^\varepsilon$. The terms $\mathcal{Q}_3$ and $\mathcal{Q}_4$ involve

$$\begin{aligned}
\mathcal{Q}_{1,1} &= \varepsilon^2 \Phi(b, u^\varepsilon_a, \varepsilon v^\varepsilon)\, v^\varepsilon\, \partial_j v^\varepsilon \partial_k b\,,\\
\mathcal{Q}_{1,2} &= \varepsilon^2 \Phi(b, u^\varepsilon_a, \varepsilon v^\varepsilon)\, v^\varepsilon\, v^\varepsilon\, \partial_j u^\varepsilon_a \partial_k b\,,\\
\mathcal{Q}_{3,1} &= \varepsilon^2 \Phi(b, u^\varepsilon_a, \varepsilon v^\varepsilon)\, v^\varepsilon\, \partial^2_{j,k} v^\varepsilon\,,\\
\mathcal{Q}_{3,2} &= \varepsilon^2 \Phi(b, u^\varepsilon_a, \varepsilon v^\varepsilon)\, \partial_k v^\varepsilon\, \partial_j v^\varepsilon\,,\\
\mathcal{Q}_{3,3} &= \varepsilon^2 \Phi(b, u^\varepsilon_a, \varepsilon v^\varepsilon)\, v^\varepsilon\, \partial_j v^\varepsilon\, \partial_k u^\varepsilon_a\,,\\
\mathcal{Q}_{4,1} &= \varepsilon^2 \Phi(b, u^\varepsilon_a, \varepsilon v^\varepsilon)\, v^\varepsilon \partial_k v^\varepsilon \partial_j u^\varepsilon_a\,,\\
\mathcal{Q}_{4,2} &= \varepsilon^2 \Phi(b, u^\varepsilon_a, \varepsilon v^\varepsilon)\, v^\varepsilon\, v^\varepsilon\, \partial_k u^\varepsilon_a \partial_j u^\varepsilon_a\,,\\
\mathcal{Q}_{4,3} &= \varepsilon^2 \Phi(b, u^\varepsilon_a, \varepsilon v^\varepsilon)\, v^\varepsilon\, v^\varepsilon\, \partial^2_{j,k} u^\varepsilon_a\,.
\end{aligned}$$

Introduce the norms

$$\|f\|_{\mathcal{Y}^{m,\varepsilon}} := \|f\|_{\mathcal{H}^m} + \varepsilon \|f\|_{L^\infty}\,, \tag{6.13}$$

$$\begin{aligned}
\|u\|_{\mathcal{X}^{m,\varepsilon}} := \|u\|_{\mathcal{H}^m} &+ \varepsilon^{1/2}\|\nabla_x u\|_{\mathcal{H}^m} + \varepsilon^{3/2}\|\nabla^2_x u\|_{\mathcal{H}^m}\\
&+ \sum_{|I|\le 2} \|Z_I u\|_{L^\infty} + \varepsilon \sum_{|I|\le 1} \|Z_I \nabla_x u\|_{L^\infty} + \varepsilon^2 \|\nabla^2_x u\|_{L^\infty}\,.
\end{aligned} \tag{6.14}$$

We denote by $\mathcal{Y}^m$ and $\mathcal{X}^m$ the natural spaces (independent of ε) associated to these norms. We denote by $\mathcal{Y}^m_0$ [resp. $\mathcal{X}^m_0$] the subspace of $v \in \mathcal{Y}^m$ [resp. $\mathcal{X}^m$] which vanish for $t < 0$ [resp. which vanish for $t < 0$ and satisfy the boundary condition $v = 0$ on $[-T_0, T_0] \times \partial\Omega$]. In section 5, we have proved that there is a constant C_0 such that for all $\varepsilon \in]0, 1]$ and $f \in \mathcal{Y}^m_0$ the problem

$$\mathcal{P}^\varepsilon v = f\,, \quad v \in \mathcal{X}^m_0 \tag{6.15}$$

has a unique solution which satisfies

$$\|v\|_{\mathcal{X}^{m,\varepsilon}} \le C_0 \|f\|_{\mathcal{Y}^{m,\varepsilon}}\,. \tag{6.16}$$

In order to use the implicit function theorem to the equation (6.12), the main step is to prove the following estimates.

Proposition 6.4. *For all $M \ge 0$, there is a constant $C(M)$ such that for all $\varepsilon \in]0, 1]$ and all v_1 and v_2 in $\mathcal{X}^m_0$, $Q^\varepsilon(v_1)$ and $Q^\varepsilon(v_2)$ belong to $\mathcal{Y}^m_0$ and*

$$\|\mathcal{Q}^\varepsilon(v_1)\|_{\mathcal{Y}^m_\varepsilon} \le \varepsilon^{1/4}\, C(M)\,, \tag{6.17}$$

$$\|\mathcal{Q}^\varepsilon(v_1) - \mathcal{Q}^\varepsilon(v_2)\|_{\mathcal{Y}^m_\varepsilon} \le \varepsilon^{1/4}\, C(M)\, \|v_1 - v_2\|_{\mathcal{X}^m_\varepsilon}\,, \tag{6.18}$$

provided that

$$\varepsilon\|v_1\|_{L^\infty} \le 1\,, \quad \varepsilon\|v_2\|_{L^\infty} \le 1\,, \tag{6.19}$$

and

$$\|v_1\|_{\mathcal{X}^{m,\varepsilon}} \le M\,, \quad \|v_2\|_{\mathcal{X}^{m,\varepsilon}} \le M\,. \tag{6.20}$$

We first investigate the L^∞ bounds. The expressions $\mathcal{Q}_\alpha$ are bilinear in v, $\varepsilon\nabla_x v$, $\varepsilon^2\nabla_x v$ with coefficients $C(b, u^\varepsilon_a, \varepsilon v)$, $\varepsilon\nabla_x b$, $\varepsilon\nabla_x u^\varepsilon_a$, and $\varepsilon^2\nabla_x u^\varepsilon_a$ which are bounded when $|\varepsilon v| \le 1$. Therefore, if v_1 and v_2 satisfy (6.19) and

$$\|v_k\|_{\mathcal{L}^\infty_\varepsilon} := \|v_k\|_{L^\infty} + \varepsilon\|\nabla_x v_k\|_{L^\infty} + \varepsilon^2\|\nabla^2_x v_k\|_{L^\infty} \le \|v_k\|_{\mathcal{X}^m_\varepsilon} \le M\,,$$

one has

$$(6.21)\qquad \begin{aligned} &\|\mathcal{Q}^\varepsilon(v_1)\|_{L^\infty} \le C(M)\,, \\ &\|\mathcal{Q}^\varepsilon(v_1) - \mathcal{Q}^\varepsilon(v_2)\|_{L^\infty} \le C(M)\,\|v_1 - v_2\|_{\mathcal{L}^\infty_\varepsilon}\,. \end{aligned}$$

Next we consider the $\mathcal{H}^m$-bounds. We use he following estimates, which follow from Gagliardo-Nirenberg inequalities.

LEMMA 6.5. *i) There is a constant C such that for all v and w in $\mathcal{H}^m \cap L^\infty$:*

$$(6.22)\qquad \|vw\|_{\mathcal{H}^m} \le C\big(\|v\|_{\mathcal{H}^m}\|w\|_{L^\infty} + \|v\|_{L^\infty}\|w\|_{\mathcal{H}^m}\big)\,.$$

ii) If $\Phi(b,u,v)$ is a smooth function of its arguments such that $\Phi(b,u,0)=0$, then for all $v \in \mathcal{H}^m \cap L^\infty)$ which satisfies $\|\varepsilon v\|_{L^\infty} \le 1$, the function $\Phi^\varepsilon(t,x) = \Phi\big(b(t,x), u^\varepsilon_a(t,x), \varepsilon v(t,x)\big)$ satisfies

$$(6.23)\qquad \|\Phi^\varepsilon\|_{\mathcal{H}^m} \le C\,\varepsilon\|v\|_{\mathcal{H}^m}\,.$$

In addition, one has the following estimates, where φ denotes the defining function of $\partial\Omega$.

LEMMA 6.6. *There is a constant C such that for all $v \in \mathcal{X}^m$ which vanishes on the boundary and all $\varepsilon' \in]0,1]$,*

$$(6.24)\qquad \|e^{-\varphi/\varepsilon'}v\|_{\mathcal{H}^m} \le C\varepsilon'\|\nabla_x v\|_{\mathcal{H}^m}\,.$$

PROOF. For ε' away from zero, or in any compact domain $\overline{\Omega}_1 \subset \Omega$ such that $e^{-\varphi/\varepsilon'} \lesssim \varepsilon'$, the estimate follows from Poincaré's inequality. Near the boundary, one can prove the estimate in local coordinates patches, and then the estimate follows from (4.70). □

PROPOSITION 6.7. *Suppose that Q is a bilinear mapping on $\mathbb{C}^N \times \mathbb{C}^N$ and $\theta > 0$ is given. There there is a constant C such that for all $\varepsilon \in]0,1]$ and all v_1 and v_2 in $\mathcal{X}^m_0$, one has*

$$(6.25)\qquad \|Q(v_1, \varepsilon\partial_j v_2)\|_{\mathcal{H}^m} \le C\,\varepsilon^{1/4}\,\|v_1\|_{\mathcal{X}^m,\varepsilon}\,\|v_2\|_{\mathcal{X}^m,\varepsilon}\,,$$

$$(6.26)\qquad \|Q(v_1, \varepsilon^2\partial^2_{j,k} v_2)\|_{\mathcal{H}^m} \le C\,\varepsilon^{1/4}\,\|v_1\|_{\mathcal{X}^m,\varepsilon}\,\|v_2\|_{\mathcal{X}^m,\varepsilon}\,,$$

$$(6.27)\qquad \|Q(\varepsilon\partial_k v_1, \varepsilon\partial_j v_2)\|_{\mathcal{H}^m} \le C\,\varepsilon^{1/2}\,\|v_1\|_{\mathcal{X}^m,\varepsilon}\,\|v_2\|_{\mathcal{X}^m,\varepsilon}\,,$$

$$(6.28)\qquad \|e^{-\theta\varphi/\varepsilon}\,Q(v_1, v_2)\|_{\mathcal{H}^m} \le C\,\varepsilon^{1/2}\,\|v_1\|_{\mathcal{X}^m,\varepsilon}\,\|v_2\|_{\mathcal{X}^m,\varepsilon}\,.$$

PROOF. The estimate (6.27) follows directly from the inequality (6.22) in Lemma 6.5, since

$$(6.29)\qquad \|\varepsilon\nabla_x v\|_{L^\infty} \le \|v\|_{\mathcal{X}^m,\varepsilon} \quad \text{and} \quad \|\varepsilon\nabla_x v\|_{\mathcal{H}^m} \le \sqrt{\varepsilon}\|v\|_{\mathcal{X}^m,\varepsilon}\,.$$

Similarly, the estimate (6.28) follows from Lemma 6.5, since, by Lemma 6.6,

$$\|e^{-\theta\varphi/2\varepsilon}v\|_{L^\infty} \le \|v\|_{\mathcal{X}^m,\varepsilon} \quad \text{and} \quad \|e^{-\theta\varphi/2\varepsilon}v\|_{\mathcal{H}^m} \lesssim \|\varepsilon\nabla_x v\|_{\mathcal{H}^m} \lesssim \sqrt{\varepsilon}\|v\|_{\mathcal{X}^m,\varepsilon}.$$

The proof of (6.25) is a little more subtle. With $\varepsilon' = \varepsilon^{3/4}$, we split the $Q(v_1, \varepsilon\partial_j v_2)$ into

$$Q(v_1, \varepsilon\partial_j v_2) = Q(e^{-\varphi/\varepsilon'}v_1, \varepsilon\partial_j v_2) + Q(v_1, (1 - e^{-\varphi/\varepsilon'})\varepsilon\partial_j v_2)$$

Using (6.29) for v_2 and the bounds

$$\|v_1'\|_{L^\infty} \le \|v_1\|_{\mathcal{X}^m,\varepsilon} \quad \text{and} \quad \|v_1'\|_{\mathcal{H}^m} \lesssim \varepsilon'\|\nabla_x v_1\|_{\mathcal{X}^m,\varepsilon} \lesssim \varepsilon^{1/4}\|v_1\|_{\mathcal{X}^m,\varepsilon}$$

for $v_1' = e^{-\varphi/\varepsilon'} v_1$, the estimate (6.22) implies that $\|Q(v_1', \varepsilon\partial_j v_2)\|_{\mathcal{H}^m}$ is bounded by the right hand side of (6.25).

On the other hand, we use that

$$\|v_1\|_{L^\infty} \le \|v_1\|_{\mathcal{X}^{m,\varepsilon}}\,, \qquad \|v_1\|_{\mathcal{H}^m} \le \|v_1\|_{\mathcal{X}^{m,\varepsilon}}\,.$$

Moreover, $w = (1 - e^{-\varphi/\varepsilon'})\varepsilon\partial_j v$ satisfies

$$\|w\|_{\mathcal{H}^m} \lesssim \|\varepsilon\partial_j v_2\|_{\mathcal{H}^m} \lesssim \sqrt{\varepsilon}\|v_2\|_{\mathcal{X}^{m,\varepsilon}}\,.$$

since, for all I, the functions $Z_I(1-\varepsilon^{-\varphi/\varepsilon'})$ are bounded on Ω uniformly in $\varepsilon' \in]0,1]$. In addition, since $1 - e^{-z} \le z$, for $z \ge 0$, we have

$$\|w\|_{L^\infty} \le \varepsilon^{1/4}\|\varphi\partial_j v_2\|_{L^\infty} \lesssim \varepsilon^{1/4} \sum_{|I|=1} \|Z_I v_2\|_{L^\infty} \lesssim \varepsilon^{1/4}\|v_2\|_{\mathcal{X}^{m,\varepsilon}}\,.$$

Thus, using (6.22), one obtains that $\|Q(v_1, w)\|_{\mathcal{H}^m}$ is bounded by the right hand side of (6.25).

The proof of (6.26) is similar. One has

$$\|\varepsilon^2\partial^2_{j,k} v_2\|_{L^\infty} \le \|v_2\|_{\mathcal{X}^{m,\varepsilon}}\,, \qquad \|\varepsilon^2\partial^2_{j,k} v_2\|_{\mathcal{H}^m} \le \varepsilon^{1/2}\|v_2\|_{\mathcal{X}^{m,\varepsilon}}$$

Hence, with $v_1' = e^{-\varphi/\varepsilon'} v_1$ as above, $\|Q(v_1', \varepsilon^2\partial_{j,k} v_2)\|_{\mathcal{H}^m}$ is bounded by the right hand side of (6.26). On the other hand, $w' = (1 - e^{-\varphi/\varepsilon'})\varepsilon^2\partial_{j,k} v$ satisfies

$$\|w'\|_{\mathcal{H}^m} \lesssim \sqrt{\varepsilon}\|v_2\|_{\mathcal{X}^{m,\varepsilon}}\,.$$

$$\|w'\|_{L^\infty} \le \varepsilon^{1/4}\|\varphi\varepsilon\partial^2_{j,k} v_2\|_{L^\infty} \lesssim \varepsilon^{1/4} \sum_{|I|=1} \|\varepsilon Z_I \nabla_x v_2\|_{L^\infty} \lesssim \varepsilon^{1/4}\|v_2\|_{\mathcal{X}^{m,\varepsilon}}\,.$$

Therefore, $\|Q(v_1, w')\|_{\mathcal{H}^m}$ is also bounded by the right hand side of (6.26). □

PROOF OF 6.4. **a)** Recall that the L^∞ bounds follow from (6.21). We prove the estimates (6.17) (6.18) with $\mathcal{H}^m$ norms in the left hand sides, for each term $\mathcal{Q}_1 \ldots \mathcal{Q}_{4,3}$. We write

$$\Phi(b, u^\varepsilon_a, \varepsilon v) = \Phi(b, u^\varepsilon_a, 0) + \Phi'(b, u^\varepsilon_a, \varepsilon v)\,,$$

where the function $\Phi'(b, u, v)$ vanishes when $v = 0$. We first consider the terms $\mathcal{Q}^0_\alpha$ obtained by replacing $\Phi(b, u^\varepsilon_a, \varepsilon v)$ by $\Phi^0(b, u^\varepsilon_a) = \Phi(b, u^\varepsilon_a, 0)$ in the definition of $\mathcal{Q}_\alpha$. Then $\mathcal{Q}^0_\alpha = \Phi^0 Q_\alpha$ with Q_α one of the following quadratic terms:

$$Q(v, \partial_j v)\,, \quad Q(\partial_k v, \partial_j v)\,, \quad Q(v, \partial^2_{j,k} v)\,, \quad Q(v, v)h^\varepsilon\,, \quad Q(v, \partial_j v)h^\varepsilon\,, \tag{6.30}$$

with h^ε either $\varepsilon\partial_k u^\varepsilon_a$, $\varepsilon\partial_k b$, $\varepsilon^2\partial_j u^\varepsilon_a\partial_k b$, $\varepsilon^2\partial_j u^\varepsilon_a\partial_k u^\varepsilon_a$ or $\varepsilon^2\partial_{j,k} u^\varepsilon_a$. In any case, we see that

$$h^\varepsilon = \varepsilon\Psi^\varepsilon + H(x, b, u_0, \varphi/\varepsilon)\,,$$

where the conormal derivatives $Z_I\Psi^\varepsilon$ are uniformly bounded for $|I| \le m$, and $H(x, b, u, z)$ is smooth and exponentially decaying in z (it is a z-derivative of the profile $\mathcal{W}$, multiplied by derivatives of φ). In particular, one can factor out a small exponential $e^{-\theta z}$ in H and write

$$h^\varepsilon = \varepsilon\Psi^\varepsilon + e^{-\theta\varphi/\varepsilon}\Psi^\varepsilon_1$$

with Ψ^ε_1 uniformly bounded with uniformly bounded conormal derivatives. The conormal derivatives of the coefficients Φ^0 are uniformly bounded and thus

$$\|\Phi^0 Q\|_{\mathcal{H}^m} \lesssim \|Q\|_{\mathcal{H}^m}\,.$$

Therefore, to prove the estimates (6.21) for $\mathcal{Q}^0_\alpha$ it is sufficient to prove them for Q_α. Thus, they directly follow from Proposition 6.7 in the first three cases. In the fifth case, it also follows from (6.25) using that the conormal derivatives of h^ε are bounded. In the fourth case, we split h^ε as indicated above and reduce the problem to estimating

$$\varepsilon Q(v_1, v_2)\,, \quad e^{-\theta\varphi/\varepsilon} Q(v_1, v_2)\,.$$

The second case also follows from Proposition 6.7 and the first case is easier: thanks to the extra factor ε, it is an immediate consequence of the estimate (6.22).

b) It remains to prove the estimates (6.21) in $\mathcal{H}^m$ norm for $\mathcal{Q}_\alpha - \mathcal{Q}^0_\alpha$ which has the form

$$\Phi'(b, u^\varepsilon_a, \varepsilon v^\varepsilon)\, Q_\alpha(v, v)\,.$$

To prove the estimates simultaneously, it is sufficient to prove that if $\Phi'(b, u, v)$ vanishes when $v = 0$, then

$$\|\Phi'(b, u^\varepsilon_a, \varepsilon v_3) Q_\alpha(v_1, v_2)\|_{\mathcal{H}^m} \lesssim \varepsilon \|v_1\|_{\mathcal{H}^{m},\varepsilon}\, \|v_2\|_{\mathcal{H}^{m},\varepsilon}\, \|v_3\|_{\mathcal{H}^{m},\varepsilon}\,. \tag{6.31}$$

We use the estimate (6.22):

$$\|\Phi' Q_\alpha\|_{\mathcal{H}^m} \lesssim \|\Phi'\|_{L^\infty} \|Q_\alpha\|_{\mathcal{H}^m} + \|\Phi'\|_{\mathcal{H}^m} \|Q_\alpha\|_{L^\infty}\,.$$

Since Φ' vanishes when $v = 0$, one has, when $\varepsilon\|v\|_{L^\infty} \le 1$,

$$\|\Phi'(b, u^\varepsilon_a, \varepsilon v_3)\|_{L^\infty} \le \varepsilon \|v_3\|_{L^\infty}$$

and, with Lemma 6.5,

$$\|\Phi'(b, u^\varepsilon_a, \varepsilon v_3)\|_{\mathcal{H}^m} \lesssim \varepsilon\, \|v_3\|_{\mathcal{H}^m}\,.$$

The $\mathcal{H}^m$ norms of Q_α are given by step a). It is sufficient here to use the weaker estimates which follow directly from (6.22) in Lemma 6.5:

$$\|Q_\alpha(v_1, v_2)\|_{L^\infty} + \|Q_\alpha(v_1, v_2)\|_{\mathcal{H}^m} \lesssim \|v_1\|_{\mathcal{X}^m,\varepsilon}\, \|v_2\|_{\mathcal{X}^m,\varepsilon}\,.$$

Thanks to the extra factor ε in the estimates of Φ', these estimate imply (6.31) and the proof of Proposition 6.4 is now complete. □

Proof of Theorem 6.1. Theorems 5.1 and 5.8 imply that $\mathcal{P}^\varepsilon$ is an isomorphism from $\mathcal{X}^m_0$ onto $\mathcal{Y}^m_0$. Thus the equation (6.12) is equivalent to

$$v^\varepsilon = (\mathcal{P}^\varepsilon)^{-1}\big(f^\varepsilon - \mathcal{Q}^\varepsilon(v)\big)\,, \quad v^\varepsilon \in \mathcal{X}^m_0\,. \tag{6.32}$$

The estimates in Theorems 5.1 5.8 and Lemma 6.2 imply that there is a constant C_1 such that for all $\varepsilon \in]0, 1]$

$$\|(\mathcal{P}^\varepsilon)^{-1} f^\varepsilon\|_{\mathcal{X}^m,\varepsilon} \le C_1\,.$$

For all $M > 0$, introduce

$$\mathcal{X}^m_0(M, \varepsilon) = \big\{ v \in \mathcal{X}^m_0\,,\ \varepsilon\|v\|_{L^\infty} \le 1 \ \text{ and } \ \|v\|_{\mathcal{X}^m,\varepsilon} \le 1 \big\}\,.$$

Moreover, Theorems 5.1 and 5.8 and Proposition (6.4) imply that for all $M > 0$, there is $C(M)$ such that for all $\varepsilon \in]0, 1]$:

$$\|(\mathcal{P}^\varepsilon)^{-1}\mathcal{Q}^\varepsilon(v)\|_{\mathcal{X}^m_\varepsilon} \le \varepsilon^{1/4}\, C(M)\,,$$

$$\|(\mathcal{P}^\varepsilon)^{-1}\mathcal{Q}^\varepsilon(v_1) - \mathcal{Q}^\varepsilon(v_2)\|_{\mathcal{X}^m_\varepsilon} \le \varepsilon^{1/4}\, C(M)\, \|v_1 - v_2\|_{\mathcal{X}^m_\varepsilon}\,,$$

for v's in $\mathcal{X}_0^m(M,\varepsilon)$. Choosing first $M > C_1$, the estimates above imply that there is $\varepsilon_0 > 0$ such that for all $\varepsilon \in]0,\varepsilon_0]$, the equation (6.32) has a unique solution in $\mathcal{X}_0^m(M,\varepsilon)$. In particular

$$\forall \varepsilon \in]0,\varepsilon_0], \quad \|v^\varepsilon\|_{\mathcal{H}^m} + \|v^\varepsilon\|_{L^\infty} \leq \|v^\varepsilon\|_{\mathcal{X}^{m,\varepsilon}} \leq M .$$

Thus we have constructed a solution $u^\varepsilon = u_a^\varepsilon + \varepsilon v^\varepsilon$ of (6.4) and $u^\varepsilon - u_0^\varepsilon = \varepsilon(u_1^\varepsilon + v^\varepsilon)$ satisfies the uniform estimates (6.5), proving Theorem 6.1. □

□

1. Appendix A. Kreiss symmetrizers

The goal of this appendix is to construct symmetrizers in the low frequency regime, as indicated in Lemma 2.13. This is an extension to our hyperbolic/parabolic setting of Kreiss' construction for hyperbolic systems (see [**Kr**] or [**Ch-P**] for another presentation). We first prove the block decomposition announced in Lemma 2.10. Next we define the extension of the spaces $\mathbb{E}_-$ to $\rho = 0$ and prove (the main part of) Proposition 1.6. Finally we proceed to the construction of the symmetrizers.

1.1. The block structure condition. Proof of Lemma 2.10. We use the notations of section 2 assuming only that Assumption 1.1 is satisfied. The matrix H is given by Lemma 2.9 and in polar coordinates $\zeta = \rho\check{\zeta}$ we write $H(p,\zeta) = \rho\check{H}(p,\check{\zeta},\rho)$ as in (2.40). We start with several remarks about the symbols of the equations. We denote here by

$$A(p,\eta,\xi) = \sum_{j<d} \eta_j A_j(p) + \xi A_d(p)$$

the symbol of the hyperbolic part of the equation and by

$$B(p,\eta,\xi) = \sum_{j,k<d} \eta_j\eta_k B_{j,k}(p) + \sum_{j<d} \xi\eta_j\big(B_{j,d}(p) + B_{d,j}(p)\big) + \xi^2 B_{d,d}(p)$$

the symbol of the parabolic part. Then

$$\det\big((i\tau+\gamma)\mathrm{Id} + iA(p,\eta,\xi) + B(p,\eta,\xi)\big) = \det\big(B_{d,d}(p)\big)\det\big(i\xi\mathrm{Id} - \mathcal{G}(p,\zeta)\big)$$

and in the polar coordinates (2.40)

$$\begin{aligned} &\det\big((i\check{\tau}+\check{\gamma})\mathrm{Id} + iA(p,\check{\eta},\check{\xi}) + \rho B(p,\check{\eta},\check{\xi})\big) = \\ &\quad \det\big(B_{d,d}(p)\big)\det\big(i\check{\xi}\mathrm{Id} - \check{H}(p,\check{\zeta},\rho)\big)\det\big(i\rho\check{\xi} - P(p,\rho\check{\zeta})\big) . \end{aligned} \tag{1.1}$$

Denote by $\lambda_j(p,\eta,\xi)$ the eigenvalues (of constant multiplicity by (H2)) of the hyperbolic symbol $A(p,\eta,\xi)$ and by $\Pi_j(p,\xi)$ the associated eigenprojectors. For ρ small, there is spectral projector $\Pi_j(p,\xi,\rho)$ of $iA(p,\eta,\xi) + \rho B(p,\eta,\xi)$ yielding a diagonal block decomposition, with $\alpha_j \times \alpha_j$ diagonal blocks

$$i\lambda_j(p,\eta,\xi)\mathrm{Id} + \rho B_j'(p,\eta,\xi,\rho).$$

where α_j is the multiplicity of λ_j. The eigenvalues of $iA(p,\xi) + \rho B(p,\xi,\rho)$ close to $i\lambda_j$ are $i\lambda_j + \rho\lambda' + O(\rho^2)$, λ' being an eigenvalue of B_j'. By (H3), one must have $\operatorname{Re}\lambda' > 0$ and therefore the spectrum of B_j' is contained in $\{\operatorname{Re}\lambda' > 0\}$. One has

$$\Delta(p,\tau,\eta,\xi,\rho) := \det\big(i\tau\mathrm{Id} + iA + \rho B\big) = \prod \Delta_j(p,\tau,\eta,\xi,\rho) \tag{1.2}$$

with

$$\Delta_j(p,\tau,\eta,\xi,\rho) := \det\big(i(\tau+\lambda_j)\mathrm{Id} + \rho B_j'(p,\eta,\xi,\rho)\big). \tag{1.3}$$

We now proceed to the proof of Lemma 2.10. Denote by H_0 the leading, first order part of H:

$$H_0(p,\check\zeta) = -(A_d^\infty(p))^{-1}\big(i\check\tau + \check\gamma)\mathrm{Id} + \sum_{j=1}^{d-1} i\check\eta_j A_j^\infty(p)\big)\,.$$

Then

$$\check H(p,\check\zeta,\rho) = H_0(p,\check\zeta) + O(\rho)\,.$$

The hyperbolicity assumption (H2) implies that the real part of the eigenvalues of H_0 do not vanish when $\check\gamma > 0$. This remains true for small ρ. Thus, when $\underline{\check\gamma} > 0$, the block reduction (2.41) holds in a neighborhood of $(\underline{p},\underline{\check\zeta},0)$ with two blocks corresponding to the eigenvalues with positive/negative real part. The first block satisfies item $i)$ in Lemma 2.10 and the second satisfies $ii)$.

Next consider the critical case that $\underline{\check\gamma} = 0$. We can perform a first smooth spectral block reduction around $(\underline{p},\underline{\check\zeta},0)$:

$$V^{-1}\check H V = \mathrm{Diag}(\mathcal{Q}_k + \rho\mathcal{R}_k) \tag{1.4}$$

which corresponds to distinct eigenvalues of $H_0(\underline{p},\underline{\check\zeta})$. The blocks corresponding to eigenvalues with positive or negative satisfy $i)$ and $ii)$ respectively.

Consider a purely imaginary eigenvalue $\underline{\mu} = i\underline{\check\xi}$ of $H_0(\underline{p},\underline{\check\zeta})$. Note that $(\underline{\check\eta},\underline{\check\xi}_d) \neq 0$ since $(\underline{\check\tau},\underline{\check\eta}) \neq 0$. We proceed by a series of steps paralleling the approach of [**Mé3**].

Since $\underline{\mu} = i\underline{\check\xi}$ is an eigenvalue of $H_0(\underline{p},\underline{\check\zeta})$, there is a unique eigenvalue λ_j of $A(p,\eta,\xi)$ such that $\underline{\check\tau} + \lambda_j(\underline{p},\underline{\check\eta},\underline{\check\xi}) = 0$. Since λ_j is real analytic in ξ, there is an integer $\nu \geq 1$ such that

$$\partial_\xi\lambda_j = \cdots = \partial_\xi^{\nu-1}\lambda_j = 0\,,\quad \partial_\xi^\nu\lambda_j = \nu!\beta \neq 0 \quad \text{at } (\underline{p},\underline{\check\eta},\underline{\check\xi}).$$

Note that β is real. Then, with $\xi' = \xi - \underline{\check\xi}$ and $\tau' = \tau - \underline{\check\tau}$, possibly complex, there holds

$$\begin{aligned}\Delta_j(\underline{p},\tau,\underline{\eta},\xi,\rho) =&\det\big((i(\tau' + \beta\xi'^\nu)Id + \rho\underline{B}'_j)\\ &+ O\big((|\rho| + |\xi'|)\,(|\tau'| + |\xi'|^\nu + \rho)^{\alpha_j}\big)\end{aligned} \tag{1.5}$$

with α_j equal to the dimension of the block, i.e. the multiplicity of λ_j, $\underline{B}'_j = B'_j(\underline{p},\underline{\check\eta},\underline{\check\xi},0)$. Indeed,

$$\big(i\tau + \lambda_j(\underline{p},\underline{\check\eta},\xi)\big)\mathrm{Id} + \rho B'_j = i(\tau' + \beta\xi'^\nu)\mathrm{Id} + \rho\underline{B}'_j + O\big(|\xi'|^{\nu+1} + \rho|\xi'| + \rho^2\big)$$

and (1.5) follows.

In the block reduction (1.4) of the boundary problem near $(\underline{p},\underline{\check\tau},\underline{\check\eta},0)$, the eigenvalue $i\underline{\check\xi}$ yields for $\rho 0$ small a block $\mathcal{Q}(p,\check\zeta) + \rho\mathcal{R}(p,\check\zeta,\rho)$. According to [**Mé3**] applied when $\rho = 0$, the constant multiplicity assumption (H2) implies that one can choose the conjugation matrix V such that

$$\mathcal{Q}(p,\check\zeta) = \begin{bmatrix} Q & \cdots & 0 \\ 0 & \ddots & 0 \\ 0 & \cdots & Q \end{bmatrix} \tag{1.6}$$

with α_j diagonal blocks all equal to the same matrix Q of size ν. Moreover, at the base point

$$Q(\underline{p},\underline{\check\zeta}) = \underline{Q} = i(\underline{\xi}Id + N_j),$$

where N_j is the Jordan's matrix of size α_j. This proves (2.44).

In addition, following [**Ral**] [**Ch-P**] one can choose the basis such that Q has the form

$$Q(p, \check{\zeta}) = \begin{bmatrix} * & 0 \dots 0 \\ \vdots & 0 \dots 0 \\ a & 0 \dots 0 \end{bmatrix} \tag{1.7}$$

and Q is purely imaginary when $\gamma = 0$ (see [**Mé3**]).

Write $\mathcal{R}$ as a block matrix, with blocks $R_{p,q}$ as in (2.43). One can perform a change of basis such that, in addition to the other properties, there holds at the base point $(\underline{p}, \underline{\check{\zeta}}, 0)$

$$R_{p,q}(\underline{p}, \underline{\check{\zeta}}, 0) = \underline{R}_{p,q} = \begin{bmatrix} * & 0 \dots 0 \\ \vdots & 0 \dots 0 \\ r_{p,q} & 0 \dots 0 \end{bmatrix}. \tag{1.8}$$

The change of basis is $Id + \rho\mathcal{T}$. Then,

$$(Id + \rho\mathcal{T})^{-1}\mathcal{Q}(Id + \rho\mathcal{T}) = \mathcal{Q} + \rho\mathcal{R} + \rho[\mathcal{Q}, \mathcal{T}] + O(\rho^2).$$

Denoting by $T_{p,q}$ the blocks of $\mathcal{T}$, at the base point $(\underline{p}, \underline{\check{\zeta}}, 0)$, the blocks of $\mathcal{R} + [\mathcal{Q}, \mathcal{T}]$ are $\underline{R}_{p,q} + [N, \underline{T}_{p,q}]$. Thus, to get (1.8), it is sufficient to choose the blocks $T_{p,q}$ such that the columns of index 2 to ν in $\underline{R}_{p,q} + [N, \underline{T}_{p,q}]$ vanish. Dropping the indices (p, q) for simplicity, this can be achieved as follows. Consider the canonical basis $(e_1, \dots, e_\nu)$ of $\mathbb{C}^\nu$. Then $Ne_1 = 0$ and $Ne_l = e_{l-1}$ for $l \geq 2$. Define T by $Te_\nu = 0$ and inductively $Te_l = NTe_{l+1} + Re_{l+1}$ for $l < \nu$. Then $[T, N]e_l = Re_l$ for $l = 2, \dots \nu$. This reduction is already used in the proof Ralston's lemma to prove that (1.7) can be achieved (see [**Ral**] and the proof of Lemma 5.4 chap 7 in [**Ch-P**]).

Comparing the eigenvalues equations (1.1) and (1.2) we see that

$$\Delta_j(p, \check{\tau} - i\check{\gamma}, \check{\eta}, \check{\xi}, \rho) = c \det\big(i\check{\xi}Id - \mathcal{Q}(p, \check{\zeta}) - \rho\mathcal{R}(p, \check{\zeta}, \rho)\big), \tag{1.9}$$

with $c \neq 0$ near the base point. We now compare the Taylor expansion (1.5) of Δ_j to the Taylor expansion of the right hand side. There we use the following lemma, in which $\mathcal{N}$ is the block diagonal matrix

$$\mathcal{N} = \begin{bmatrix} N_j & \dots & 0 \\ 0 & \ddots & 0 \\ 0 & \dots & N_j \end{bmatrix}.$$

LEMMA 1.1. *Suppose that $\mathcal{M}(h)$ is a $\alpha_j\nu \times \alpha_j\nu$ matrix with blocks $M_{p,q}(h)$ depending smoothly on the parameter h, satisfying* (1.8) *and such that $\mathcal{M}(0) = 0$. Then there holds*

$$\det\big(\xi Id - \mathcal{N} + i\mathcal{M}(h)\big) = \det(\xi^\nu Id + ih\partial_h M^\flat(0)) + O\big((|h| + |\xi|)\,(|\xi|^\nu + |h|)^{\alpha_j}\big),$$

where $M^\flat$ is the $\alpha_j \times \alpha_j$ matrix with entries $m_{p,q}$ which are the lower left hand corner coefficient of $M_{p,q}$.

We apply this lemma first with $h = \gamma$ and $\mathcal{M}(h) = \mathcal{Q}(\underline{p}, \underline{\check{\tau}}, \underline{\check{\eta}}, \gamma) - \underline{\mathcal{Q}}$. Then $M^\flat = a\text{Id}$ where a is the lower left hand corner coefficient of Q as in (1.7). Then with $\xi' = \check{\xi} - \check{\underline{\xi}}$,

$$\begin{aligned}\det\Big(i\xi - \mathcal{Q}(\underline{p}, \check{\underline{\tau}}, \check{\underline{\eta}}, \gamma)\Big) &= \det\big(i\xi'\mathrm{Id} - i\mathcal{N} - \mathcal{M}\big)\\ &= i^{\nu\alpha_j}\det\big(\xi'\mathrm{Id} - \mathcal{N} + i\mathcal{M}\big)\\ &= i^{\nu\alpha_j}\Big(\xi'^{\nu} + i\gamma\partial_\gamma a(\underline{p}, \check{\underline{\zeta}})\Big)^{\alpha_j} + h.o.t.\end{aligned}$$

where $h.o.t.$ stands for higher order terms which are $O\big(\gamma + |\xi'|)(|\xi'|^{\nu} + \gamma)^{\alpha_j}\big)$. On the other hand,

$$\Delta_j(\underline{p}, \check{\underline{\tau}} - i\gamma, \check{\underline{\eta}}, \check{\xi}) = i^{\alpha_j}\big(i\gamma + \beta\xi'^{\nu})\big)^{\alpha_j} + h.o.t.$$

Thus

$$i\gamma + \beta\xi'^{\nu} = \underline{c}i^{(\nu-1)\alpha_j}\big(\xi'^{\nu} + i\gamma\partial_\gamma a(\underline{p}, \check{\underline{\zeta}})\big)$$

and

$$\partial_\gamma a(\underline{p}, \check{\underline{\zeta}}) = \beta^{-1}.$$

Since a is homomorphic in $\tau - i\gamma$ and purely imaginary when $\gamma = 0$, we already knew that $\partial_\gamma a$ is real when $\gamma = 0$. Thus, we have

$$\partial_\gamma a(\underline{p}, \check{\underline{\zeta}}) = \partial_\gamma \mathrm{Re}\, a(\underline{p}, \check{\underline{\zeta}}) = \beta^{-1}. \tag{1.10}$$

In particular, since $\beta \neq 0$, we recover here that $\partial_\gamma a(\underline{p}, \check{\underline{\zeta}}) \neq 0$ as already shown in [**Kr**] [**Ch-P**] [**Mé3**]. With (1.10), we will be able to discuss its sign.

Next, we make a second application of Lemma 1.1 with parameter $h = \rho$ and $\mathcal{M} = \rho\mathcal{R}(\underline{p}, \check{\underline{\zeta}}, \rho)$. Thus

$$\begin{aligned}&\det\big(i\xi Id - \underline{\mathcal{Q}} - \rho\mathcal{R}(\underline{p}, \check{\underline{\zeta}}, \rho)\big)\\ &\qquad = i^{\nu\alpha_j}\det\big(\xi'^{\nu} Id + i\rho\underline{R}^{\flat}\big) + h.o.t.\end{aligned} \tag{1.11}$$

where $R^{\flat}$ is the $\alpha_j \times \alpha_j$ matrix with entries $r_{p,q}$ and $\underline{R}^{\flat}$ its value at the base point. Thus, comparing the Taylor expansions (1.5) and (1.11), we find that

$$\det\big(\beta\xi'^{\nu}\mathrm{Id} + i\rho\underline{B}'_j\big) = \underline{c}i^{(\nu-1)\alpha_j}\det\big(\xi'^{\nu}\mathrm{Id} + i\underline{R}^{\flat}\big).$$

Therefore, the eigenvalues of $\underline{R}^{\flat}$ are the eigenvalues of $\beta^{-1}\underline{B}'_j$. In particular,

$$\mathrm{Spectrum}\big(\beta\underline{R}^{\flat}\big) \subset \{\mathrm{Re}\,\lambda' > 0\}. \tag{1.12}$$

With (1.10), we see that the real part of the spectrum of $\underline{R}^{\flat}$ has the same sign as $\partial_\gamma a(\underline{p}, \check{\underline{\zeta}})$, which we call κ.

From (1.12), it is a standard fact that there is a basis such that $\mathrm{Re}\,(\kappa R^{\flat})$ is positive definite, as claimed in Lemma 2.10. Thus, let T be a $\alpha_j \times \alpha_j$ matrix such that $\mathrm{Re}\, T^{-1}\kappa\underline{R}^{\flat}T$ is positive definite. Consider $\mathcal{T}$ the $\nu\alpha_j \times \nu\alpha_j$ matrix with $\nu \times \nu$ blocks $T_{p,q} = t_{p,q}\mathrm{Id}$ where $t_{p,q}$ are the coefficients of T. Then $\mathcal{S} = \mathcal{T}^{-1}$ has blocks $S_{p,q} = s_{p,q}\mathrm{Id}$ where the $s_{p,q}$ are the entries of $S = T^{-1}$. Straightforward computations show that

$$\mathcal{T}^{-1}\mathcal{Q}\mathcal{T} = \mathcal{Q},$$

since the blocks of the first matrix in the left hand side are

$$\sum_n t^{-1}_{p,n} Q t_{n,q} = Q\delta_{p,q}.$$

Next, the blocks of $\widetilde{\mathcal{R}} := \mathcal{T}^{-1}\mathcal{R}\mathcal{T}$ are

$$\widetilde{R}_{p,q} = \sum_{n,m} s_{p,n} R_{n,m} t_{m,q} \,. \tag{1.13}$$

At the base point $(\underline{p}, \check{\zeta}, 0)$ the columns 2 to ν of the $\underline{R}_{n,m}$ vanish. Since the $s_{p,n}$ and $t_{m,q}$ are scalar, the same property holds for $\underline{\widetilde{R}}_{p,q}$. Therefore, the form of the matrix $\widetilde{\mathcal{R}}$ at the base point is unchanged. Moreover, (1.13) implies that the matrix of lower left hand corner elements in $\widetilde{\mathcal{R}}$ is $\widetilde{R}^\flat = T^{-1}R^\flat T$ and thus $\kappa(\widetilde{R}^\flat + (\widetilde{R}^\flat)^*)$ is positive definite at the base point.

This finishes the proof that one can chose a basis such that the blocks $\mathcal{Q} + \rho\mathcal{R}$ associated to eigenvalues $\underline{\nu}$ which are purely imaginary satisfies the properties *iv)* listed in Lemma 2.10.

Note. We have given an argument valid for either of cases *iii)* or *iv)*. When $\nu = 1$ (nonglancing modes), case *iii)*, the construction above is much simpler, with "blocks" Q of dimension 1, and the matrix $R^\flat = R$.

Proof of Lemma 1.1

a) We start with a general remark. Consider a $N \times N$ matrix A with entries $a_{j,k}$ depending on variables x. Assume that

$$a_{j,k}(x) = \underline{a}_{j,k}(x) + h.o.t.$$

where $\underline{a}_{j,k}$ is homogeneous of degree $\mu_j - \nu_k$ and $h.o.t$ means something of higher degree, here $O(|x|^{\mu_j - \nu_k + 1})$. Then

$$\det A(x) = \det \underline{A}(x) + h.o.t. \tag{1.14}$$

and $\det \underline{A}$ is homogeneous of degree $\mu := \sum \mu_j - \sum \nu_k$. Indeed,

$$\det A = \sum \epsilon(\sigma) a_{\sigma_1,1} \cdots a_{\sigma_N,N}$$

where the sum is extended over all the permutations σ of $\{1, \ldots, N\}$ and $\epsilon(\sigma)$ is the signature of σ. Each monomial is equal to the corresponding one with $\underline{a}$ in place of a plus higher order terms, and the term with the $\underline{a}$ is homogeneous of degree

$$\sum (\mu_{\sigma_k} - \nu_k) = \sum \mu_{\sigma_k} - \sum \nu_k = \sum \mu_j - \sum \nu_k = \mu.$$

b) In our case, we consider the matrix $\mathcal{A} = \xi Id - \mathcal{N} + i\rho\mathcal{M}$. Be denote by $A_{p,q}$ the blocks in $\mathcal{A}$ and by $A_{p,a,p,b}$ the entries of $A_{p,q}$. Remember that $1 \le p, q \le \alpha$ and $1 \le a, b \le \nu$. We use a quasi-homogeneous version of (1.14). We consider using weight 1 on the variable ξ and weight ν on the variable ρ. To be more specific, with ξ_0 and h_0 fixed, consider $\xi = t\xi_0$ and $h = t^\nu h_0$ with $t \in [0, 1]$. Introduce the weights

$$\mu_{p,a} = a + 1 \,, \quad \nu_{q,b} = b \,.$$

The diagonal terms in $\mathcal{A}$ are equal to ξ, homogeneous of degree $1 = \mu_{p,a} - \nu_{p,a}$ in t. The entries $N_{p,a,q,b}$ of $\mathcal{N}$ are zero or equal to -1 when $p = q$ and $b = a+1$ which is homogeneous of degree $0 = \mu_{p,a} - \nu_{p,a+1}$. Introduce $\underline{\mathcal{M}} = \partial_h \mathcal{M}(0)$. Then the form (1.8) of $\mathcal{M}$ implies that $M_{p,a,q,b}(th)$ vanishes when $b > 1$. When $b = 1$

$$M_{p,a,q,1}(th) = t^\nu h_0 \underline{M}_{p,a,q,b} + O(t^{2\nu}) \,.$$

The leading term is homogeneous of degree ν which is strictly larger than $\mu_{p,a} - \nu_{q,1} = a$ if $a < \nu$, and exactly equal to $\mu_{p,a} - \nu_{q,1} = \nu$ if $a = \nu$. Thus, only the lower left hand corners of $M_{p,q}$ have a non vanishing principal part in the sense of a). Thus

$$\begin{aligned}(1.15)\qquad \det\big(t\xi_0 Id - \mathcal{N} + i\mathcal{M}(t^\nu h_0)\big) &= \\ &\det\big(t\xi Id - \mathcal{N} + it^\nu h_0 \underline{\mathcal{M}}^\flat\big) + O(t^{\alpha\nu+1})\end{aligned}$$

where the leading term is homogeneous in t of degree $\alpha\nu$ and $\underline{\mathcal{M}}^\flat$ is the matrix with all entries equal to zero except $\underline{M}^\flat_{p,\nu,q,1} = \underline{m}_{p,q}$.

c) Grouping the indices the other way, i.e. considering the matrix $\mathcal{A}$ as a the block matrix with blocks $\hat{A}_{a,b}$ with entries $\hat{A}_{p,a,q,b}$, we see that there is a permutation matrix P such that

$$P^{-1}\big(-\mathcal{N} + h\underline{\mathcal{M}}^\flat\big)P = \begin{bmatrix} 0 & -Id & 0 & \cdots \\ & & & \\ 0 & 0 & \ddots & 0 \\ 0 & \cdots & 0 & -Id \\ h\underline{M}^\flat & 0 & \cdots & 0 \end{bmatrix} := \widehat{\mathcal{M}}$$

where $\underline{M}^\flat$ is the matrix with entries $\underline{m}_{p,q}$. Thus $u \in \ker(\xi \mathrm{Id} - \mathcal{N} + h\underline{\mathcal{M}}^\flat)$ if and only if $v = P^{-1}u \in \ker(\xi\mathrm{Id} + \widehat{\mathcal{M}})$, which means that the blocks components v_a of v satisfy $v_a = \xi'^{a-1} v_1$ and $v_1 \in \ker(h\underline{M}^\flat + \xi'^{\nu-1} Id)$. Therefore

$$\det\big(\xi Id\mathcal{N} + ih\underline{\mathcal{M}}^\flat\big) = \det\big(\xi^\nu Id + h\underline{M}^\flat\big).$$

With (1.15) this implies that

$$\begin{aligned}(1.16)\qquad \det\big(t\xi_0 Id - \mathcal{N} + i\mathcal{M}(t^\nu h_0)\big) &= \\ &\det\big((t\xi_0)^\nu Id + it^\nu h_0 \underline{M}^\flat\big) + O(t^{\alpha\nu+1})\end{aligned}$$

and the Lemma follows. □

1.2. Stability for low frequencies. Proof of Proposition 1.6. We now assume that Assumptions 1.1 and 1.2 are satisfied. By Lemma 2.6 and (2.35), for $\zeta \neq 0$, the spaces $\mathbb{E}_-(p,\zeta)$ are related to the spaces $\mathbb{F}_-(p,\zeta)$ generated by the generalized eigenspaces of $\mathcal{G}^\infty(p,\zeta)$ associated to eigenvalues in $\{Re\mu < 0\}$:

$$\mathbb{E}_-(p,\zeta) = \mathcal{W}(0,p,\zeta)\mathbb{F}_-(p,\zeta).$$

Since the mapping $\mathcal{W}$ is smooth up to $\zeta = 0$, the limits of $\mathbb{E}_-$ as $\zeta \to 0$ are related to the limits of $\mathbb{F}_-$. By Lemma 2.5, $\mathbb{F}_-$ and thus $\mathbb{E}_-$ have dimension N and depends smoothly on ζ for $\zeta \neq 0$.

Consider polar coordinates $\zeta = \rho\check{\zeta}$, with $|\check{\zeta}| = 0$. We use the notations

$$\check{\mathbb{E}}_-(p,\check{\zeta},\rho) = \mathbb{E}_-(p,\rho,\check{\zeta}), \quad \check{\mathbb{F}}_-(p,\check{\zeta},\rho) = \mathbb{F}_-(p,\rho,\check{\zeta}).$$

They are defined for $\rho > 0$.

LEMMA 1.2. *Under Assumptions* 1.1 *and* 1.2

i) the vector bundles $\check{\mathbb{E}}_-$ *and* $\check{\mathbb{F}}_-$ *have* C^∞ *extensions up to* $\rho = 0$ *near points where* $\check{\gamma} > 0$. *We denote them by* $\mathbb{E}^0_-(p,\check{\zeta})$ *and* $\mathbb{F}^0_-(p,\check{\zeta})$;

ii) the vector bundles $\mathbb{E}^0_-$ *and* $\mathbb{F}^0_-$ *have continuous extensions to* $\check{\gamma} = 0$.

PROOF. **a)** By Lemma 2.9, $\mathcal{G}^\infty$ is conjugated to the block diagonal matrix $\mathcal{G}_2$ (see (2.37)). Thus, for $\zeta \neq 0$ small

$$(1.17)\qquad \mathbb{F}_-(p,\zeta) = \mathcal{V}(p,\zeta)\mathbb{G}_-(p,\zeta)$$

where $\mathbb{G}_-$ is the space generated by the generalized eigenspaces of $\mathcal{G}_2(p,\zeta)$ associated to eigenvalues in $\{Re\mu < 0\}$. In addition, in the block decomposition (2.37),

$$\mathbb{G}_-(p,\zeta) = \mathbb{G}_-^H(p,\zeta) \oplus \mathbb{G}_-^P(p,\zeta) \tag{1.18}$$

where the spaces $\mathbb{G}_-^H$ and $\mathbb{G}_-^P$ are associated respectively to H and P. Since $P(p,\zeta)$ has no purely imaginary eigenvalue, it follows that $\mathbb{G}_-^P(p,\zeta)$ is smooth in a neighborhood of $\zeta = 0$. Therefore, it is sufficient to study the spaces $\mathbb{G}_-^H(p,\zeta)$.

There we use the polar coordinates $\zeta = \rho\check{\zeta}$, $H = \rho\check{H}$, and we have

$$\mathbb{G}_-^H(p,\zeta) = \mathbb{H}_-(p,\zeta,\rho) \tag{1.19}$$

where $\mathbb{H}_-(p,\check{\zeta},\rho)$ is associated to $\check{H}(p,\check{\zeta},\rho)$.

Suppose that $\check{\gamma} > 0$. Then (H2) implies that $H_0(p,\check{\zeta}) = \check{H}(p,\check{\zeta},0)$ has no eigenvalues on the imaginary axis. This remains true for ρ small, and therefore $\check{\mathbb{F}}_-(p,\check{\zeta},\rho)$ is a C^∞ vector bundle for p in a neighborhood of $\underline{p}$, and $(\check{\zeta},\rho) \in \mathbb{R}^{d+2}$ with $|\check{\zeta}| = 0$, $\check{\gamma} \geq 0$, $\rho \geq 0$ and $\check{\gamma}\rho > 0$. In particular, $\mathbb{H}_-(p,\check{\zeta},0)$ is well defined for $\check{\gamma} > 0$.

Tracing back, we see that this defines $\check{\mathbb{F}}(p,\check{\zeta},\rho)$ as a C^∞ vector bundle for $\check{\gamma} > 0$ and $\rho \geq 0$, and

$$\check{\mathbb{F}}_-(p,\check{\zeta},0) = \mathcal{V}(p,0)\big(\mathbb{H}_-(p,\check{\zeta},0) \oplus \mathbb{G}_-^P(p,0)\big) \qquad \text{for } \check{\gamma} > 0\,. \tag{1.20}$$

This is transported to $\mathbb{E}_-$ using 2.35, proving that $\check{\mathbb{E}}(p,\check{\zeta},\rho)$ as a C^∞ vector bundle for $\check{\gamma} > 0$ and $\rho \geq 0$ with

$$\check{\mathbb{E}}_-(p,\check{\zeta},0) = \mathcal{W}(0,p,0)\check{\mathbb{F}}_-(p,\check{\zeta},0) \qquad \text{for } \check{\gamma} > 0\,. \tag{1.21}$$

This proves *i*).

b) To prove *ii*) it is sufficient to show that $\mathbb{H}_-(p,\check{\zeta},0)$ extends continuously to $\check{\gamma} = 0$. We can argue locally, and work around $\underline{p}$ and $\underline{\check{\zeta}}$.

In a small neighborhood of $(\underline{p},\underline{\check{\zeta}})$, we use the block decomposition (1.4) of $\check{H}$ given by Lemma 2.10. There, one has

$$\mathbb{H}_-(p,\check{\zeta},\rho) = \oplus\mathbb{H}_-^k(p,\check{\zeta},\rho) \tag{1.22}$$

where the $\mathbb{H}_-^k$ are associated to $\mathcal{Q}_k + \rho\mathcal{R}_k$. Recall that $\mathcal{Q}_k = \mathrm{Diag}(Q_k)$ as in (1.6) or (2.43). We investigate the different possibilities.

i) If the spectrum of $Q_k(p,\check{\zeta})$ is contained in $\{\mathrm{Re}\,\mu > 0\}$, this is true for $\mathcal{Q}_k$ and remains true for $\mathcal{Q}_k + \rho\mathcal{R}_k$. Thus, for ρ small

$$\mathbb{H}_-^k(p,\check{\zeta},\rho) = \{0\}\,. \tag{1.23}$$

ii) Similarly, if the spectrum of $Q_k(p,\check{\zeta})$ is contained in $\{\mathrm{Re}\,\mu < 0\}$, then, for ρ small

$$\mathbb{H}_-^k(p,\check{\zeta},\rho) = \mathbb{C}^{N_k}\,. \tag{1.24}$$

where N_k is the dimension of the k-th block $\mathcal{Q}_k$.

Only these two cases can occur when $\underline{\check{\gamma}} > 0$, and we recover here that $\mathbb{H}_-$ is smooth up to $\rho = 0$. Suppose now that $\underline{\check{\gamma}} = 0$. We consider next the other two possibilities.

iii) Suppose now that the blocks Q_k have dimension $\nu_k = 1$. Then $Q_k(p,\check{\zeta})$ is a complex number, purely imaginary when $\check{\gamma} = 0$ and $\partial_{\check{\gamma}}\mathrm{Re}\,Q_k \neq 0$ on the given neighborhood of $(\underline{p},\underline{\check{\zeta}})$. If it is positive, then $\mathrm{Re}\,Q_k > 0$ when $\check{\gamma} > 0$. In addition, we have that $\mathcal{R}_k$ is positive definite. Thus the spectrum of $\mathcal{Q}_k + \rho\mathcal{R}_k = Q_k\mathrm{Id} + \rho\mathcal{R}_k$

is contained in $\{\operatorname{Re}\mu > 0\}$. Similarly, if $\partial_{\check\gamma} Q_k < 0$, the spectrum of $\mathcal{Q}_k + \rho\mathcal{R}_k$ is contained in $\{\operatorname{Re}\mu < 0\}$. Thus

$$\mathbb{H}^k_-(p,\check\zeta,\rho) = \{0\} \qquad \text{if} \quad \partial_{\check\gamma} Q_k > 0\,, \tag{1.25}$$

$$\mathbb{H}^k_-(p,\check\zeta,\rho) = \mathbb{C}^{N_k} \qquad \text{if} \quad \partial_{\check\gamma} Q_k < 0\,. \tag{1.26}$$

In the cases i), ii) and iii) the formulas (1.23) to (1.26) have clear C^∞ extension to $\rho = 0$ and $\check\gamma = 0$.

iv) This is the most delicate case of *glancing modes.* Our goal is to continuously extend to $\check\gamma = 0$ the bundle $\mathbb{H}^k_-(p,\check\zeta,0)$. This space is associated to the operator $\mathcal{Q}_k$. In the block decomposition (2.43) $\mathcal{Q}_k = \operatorname{Diag}(Q_k)$, we see that for $\check\gamma > 0$

$$\mathbb{H}^k_-(p,\check\zeta,0) = \mathrm{H}^k_-(p,\check\zeta) \oplus \cdots \oplus \mathrm{H}^k_-(p,\check\zeta)$$

where H^k_- is the negative space associated to Q_k and the sum has α_k terms, α_k being the number of blocks Q_k in $\mathcal{Q}_k$. Recall that, by (H2), Q_k has no purely imaginary eigenvalues when $\check\gamma > 0$, so that the spaces H^k_- are well defined for $\check\gamma > 0$.

From (2.44), Q_k is a perturbation of the matrix $i(\mu_k\mathrm{Id} + N_k)$. We are now in the classical situation met in the analysis of strictly hyperbolic boundary value problems. It is known (see, e.g. [**Kr**],[**Ch-P**], [**ZS**], [**Z**]) that the subspaces H^k_- have well-defined limits when $(p,\check\zeta) \to (\underline{p},\underline{\check\zeta})$. In addition (see e.g. Remark 3.6 and Proposition 3.7, chap 7 in [**Ch-P**])

$$\mathrm{H}^k_-(\underline{p},\underline{\check\zeta}) = \mathbb{C}^{\beta_k} \times \{0\}^{\nu_k - \beta_k} \tag{1.27}$$

the space generated by the first β_k elements of the canonical basis in $\mathbb{C}^{\nu_k}$ where

$$\beta_k = \begin{cases} \frac{1}{2}\nu_k & \text{when } \nu_k \text{ is even}\,, \\ \frac{1}{2}(\nu_k \pm 1) & \text{when } \nu_k \text{ is odd} \quad \text{and } \mp\partial_{\check\gamma}\operatorname{Re} a_k > 0\,. \end{cases} \tag{1.28}$$

Recall that a_k is the lower left hand corner entry of Q_k.

Adding the different blocks, this shows that $\mathbb{H}^k_-(p,\check\zeta,0)$ and thus $\mathbb{H}_-(p,\check\zeta,0)$ have limits when $(p,\check\zeta)$ with $\check\gamma > 0$ converge to $(\underline{p},\underline{\check\zeta})$. Since $(\underline{p},\underline{\check\zeta})$ is arbitrary, the limit is defined for all p and $\check\zeta$ with $\check\zeta = 0$. As in [**Ch-P**], one can show that the bundle $\mathbb{H}_-$ is continuous also on $\check\gamma = 0$. and this finishes the proof of the lemma. □

From the proof above, we see that the bundles on $\rho = 0$ are linked by the identities:

$$\begin{aligned} &\mathbb{E}^0_-(p,\check\zeta) = \mathcal{W}(0,p,0)\mathbb{F}^0_-(p,\check\zeta)\,, \quad \mathbb{F}^0_-(p,\check\zeta) = \mathcal{V}(p,0)\mathbb{G}^0_-(p,\check\zeta), \\ &\mathbb{G}^0_-(p,\check\zeta) = \mathbb{H}_-(p,\check\zeta,0) \oplus \mathbb{G}^P_-(p,0)\,. \end{aligned} \tag{1.29}$$

We also supplement the discussion above with the following definition. We consider a point $(\underline{p};\underline{\check\zeta})$ with $|\underline{\check\zeta}| = 1$ and $\check\gamma \geq 0$.

- If the block $\mathcal{Q}_k(\underline{p},\underline{\check\zeta})$ satisfies property i) in Lemma 2.10, we define

$$\mathbb{H}^k_+(\underline{p},\underline{\check\zeta},0) = \mathbb{C}^{N_k}\,.$$

where N_k is the dimension of the k-th block $\mathcal{Q}_k$.

- If the block $\mathcal{Q}_k(\underline{p},\underline{\check\zeta})$ satisfies property ii), then:

$$\mathbb{H}^k_+(\underline{p},\underline{\check\zeta},\rho) = \{0\}\,.$$

• If $\underline{\check{\gamma}} = 0$ and the block $\mathcal{Q}_k(\underline{p}, \underline{\check{\zeta}})$ satisfies property *iii*), then we define

$$\begin{aligned} \mathbb{H}_+^k(\underline{p}, \underline{\check{\zeta}}, 0) &= \mathbb{C}^{N_k} \qquad \text{if} \quad \partial_{\check{\gamma}} Q_k > 0\,, \\ \mathbb{H}_+^k(\underline{p}, \underline{\check{\zeta}}, 0) &= \{0\} \qquad \text{if} \quad \partial_{\check{\gamma}} Q_k < 0\,. \end{aligned}$$

In the three cases above the spaces $\mathbb{H}_+$ are associated to the eigenvalues of $\mathcal{Q}_k$ with positive real parts. For glancing modes, this construction fails since the limits of positive and negative spaces intersect when $\check{\gamma} = 0$. Instead, we choose the supplementary space as in [**Kr**].

• If $\underline{\check{\gamma}} = 0$ and the block $\mathcal{Q}_k(\underline{p}, \underline{\check{\zeta}})$ satisfies property *iii*), we first define

$$\mathrm{H}_+^k(\underline{p}, \underline{\check{\zeta}}) = \{0\}^{\beta_k} \times \mathbb{C}^{\nu_k - \beta_k}$$

where β_k is defined in (1.28). In the block decomposition $\mathcal{Q}_k = \mathrm{Diag}(Q_k)$, we next define

$$\mathbb{H}_+^k \underline{p}, \underline{\check{\zeta}}, 0) = \mathrm{H}_+^k(\underline{p}, \underline{\check{\zeta}}) \oplus \cdots \oplus \mathrm{H}_+^k(\underline{p}, \underline{\check{\zeta}})$$

where the sum has α_k terms.

Adding up, we next introduce

$$\mathbb{G}_+^0(\underline{p}, \underline{\check{\zeta}}) = \mathbb{H}_+(\underline{p}, \underline{\check{\zeta}}, 0) \oplus \mathbb{G}_+^P(\underline{p}, 0) \tag{1.30}$$

where $\mathbb{G}_+^P(p, \zeta)$ is the space spanned by the eigenvectors of $P(p, \zeta)$ with eigenvalues in $\{\mathrm{Re}\,\mu > 0\}$.

Clearly, there holds

$$\begin{aligned} &\mathbb{G}_-^P(\underline{p}, 0) \oplus \mathbb{G}_+^P(\underline{p}, 0) = \mathbb{C}^N\,, \\ &\mathbb{H}_-(\underline{p}, \underline{\check{\zeta}}, 0) \oplus \mathbb{H}_+(\underline{p}, \underline{\check{\zeta}}, 0) = \mathbb{C}^N\,, \\ &\mathbb{G}_-^0(\underline{p}, \underline{\check{\zeta}}) \oplus \mathbb{G}_+^0(\underline{p}, \underline{\check{\zeta}}) = \mathbb{C}^{2N}\,. \end{aligned} \tag{1.31}$$

We denote by $\Pi_\pm^P(\underline{p})$, $\Pi_\pm^H(\underline{p}, \underline{\check{\zeta}})$ and $\Pi_\pm(\underline{p}, \underline{\check{\zeta}}) = \Pi_\pm^H \oplus \Pi_\pm^P$ the projectors associated to these decompositions.

Recall that $\Gamma_1(p, \zeta) = \Gamma \mathcal{W}^{-1}(0, p, \zeta)$ is the boundary operator deduced from Γ through the substitution $U(z) = \mathcal{W}(z, p, \zeta) U_1(z)$. We will also use the substitution $U_1(z) = \mathcal{V}(p, \zeta) U_2(z)$ and the corresponding boundary condition is $\Gamma_2(p, \zeta) = \mathcal{V}(p, \zeta)\Gamma_1(p, \zeta)$.

PROPOSITION 1.3. *In addition to Assumptions* 1.1 *and* 1.2, *suppose that Assumption* 1.4 *holds. Consider* $(\underline{p}, \underline{\check{\zeta}})$ *with* $|\check{\zeta}| = 1$ *and* $\check{\gamma} \geq 0$. *Then,* $\mathbb{E}_-^0(\underline{p}, \underline{\check{\zeta}})$, $\mathbb{F}_-^0(\underline{p}, \underline{\check{\zeta}})$ *and* $\mathbb{G}_-^0(\underline{p}, \underline{\zeta})$ *are transverse to* $\ker \Gamma$, $\ker \Gamma_1(\underline{p}, 0)$ *and* $\ker \Gamma_2(\underline{p}, 0)$ *respectively. In particular, there is* C *such that*

$$\forall V \in \mathbb{C}^{2N}: \qquad |\Pi_-(\underline{p}, \underline{\check{\zeta}})V| \leq C\Big(|\Gamma_2(\underline{p}, 0)V| + |\Pi_+(\underline{p}, \underline{\check{\zeta}})V|\Big)\,. \tag{1.32}$$

PROOF. If $\mathbb{E}$ is a N-dimensional space transversal to $\ker \Gamma$, there is an $N \times N$ matrix such that $\mathbb{E} = \{(u, v) \in \mathbb{C}^N \times \mathbb{C}^N; v = Av\}$. In this case, an orthonormal basis in $\mathbb{E}$ is obtained as the image of the canonical basis in $\mathbb{C}^N$ by $u \mapsto (Ou, AOu)$ with $O = (\mathrm{Id} + A^*A)^{-1/2}$. Thus $\det(\mathbb{E}, \ker \Gamma) = \det(\mathrm{Id} + A^*A)^{-1/2}$. This shows that if $\det(\mathbb{E}, \ker \Gamma)$ is bounded from below by a positive constant, then A is bounded. Therefore, Assumption 1.4 implies that there is C such that for all ζ with $0 < |\zeta| \leq 1$ and $\gamma \geq 0$, one has

$$\forall U \in \mathbb{E}_-(\underline{p}, \zeta): \qquad |U| \leq C|\Gamma U|\,.$$

Thus for all $\rho \in]0,1]$, $\check\zeta$ such that $|\check\zeta| = 1$ and $\check\gamma > 0$,

$$\forall U \in \check{\mathbb{E}}_-(\underline{p}, \check\zeta, \rho) : \qquad |U| \leq C|\Gamma U| .$$

By continuity, this extends to $\rho = 0$, proving that for all$\check\zeta$ such that $|\check\zeta| = 1$ and $\check\gamma > 0$, there holds

$$\forall U \in \mathbb{E}^0_-(\underline{p}, \check\zeta) : \qquad |U| \leq C|\Gamma U| ,$$

with the same constant C. By continuity, this extends to $\check\gamma = 0$. This implies that $\mathbb{E}^0(\underline{p}, \check\zeta)$ is transverse to $\ker \Gamma$. This is transported to $\mathbb{F}^0_-$ and $\mathbb{G}^0_-$ using $\mathcal{W}(0, \underline{p}, 0)^{-1}$ and $\mathcal{V}(\underline{p}, 0)^{-1}$. The estimate (1.32) follows since $\mathbb{G}^0_-(\underline{p}, \check\zeta) = \ker \Pi_+(\underline{p}, \underline{\check\zeta})$. ☐

The next result establishes the main assertion (for our purposes) of Proposition 1.6.

Proposition 1.4. *The hyperbolic boundary value problem* (1.1) (1.8) *satisfies the uniform Kreiss-Lopatinski stability condition.*

Proof. The Fourier Laplace transform at frequency $\check\zeta$ of the frozen coefficient linearized hyperbolic problem at $(\underline{b}, \underline{u})$ reads

$$\partial_z u - H_0(\underline{p}, \zeta) u = f , \quad u(0) \in T_{\underline{u}} \mathcal{C}_{\underline{b}} ,$$

with $(\underline{p} = \underline{b}, \underline{u}, 0)$ as in section 2. The spaces generated by eigenfunctions of H_0 associated to eigenvalues in $\{\mathrm{Re}\, \mu < 0\}$ are precisely $\mathbb{H}_-(\underline{p}, \check\zeta, 0)$. Thus, the uniform Kreiss-Lopatinski condition for (1.1) (1.8) reads: for all $\underline{p}$ and $\underline{\check\zeta}$ with $|\check\zeta| = 1$ and $\check\gamma \geq 0$, one has

(1.33)
i) the dimension of $\mathbb{H}_-(\underline{p}, \underline{\check\zeta}, 0)$ *is equal to* $N - N_-$,
ii) $\mathbb{H}_-(\underline{p}, \underline{\check\zeta}, 0)$ *and* $T_{\underline{u}} \mathcal{C}_{\underline{b}}$ *are transverse.*

Conditions (i)–(ii) of Definition 1.3 imply that $T_{\underline{u}} \mathcal{C}_{\underline{b}}$ is the set of end state values of solutions to the linearized equation (1.9). Written as a first order system, this equation reads

$$\partial_z U - \mathcal{G}(z, 0) U = 0, \qquad \Gamma U(0) = 0 .$$

Through the change of unknowns $U(z) = \mathcal{W}(z, \underline{p}, 0) \mathcal{V}(\underline{p}, 0) U_2(z)$, the equation is transformed into

$$\partial_z U_2 - \begin{bmatrix} 0 & \mathrm{Id} \\ 0 & P(\underline{p}, 0) \end{bmatrix} U_2 = 0, \qquad \Gamma_2(\underline{p}, 0) U_2(0) = 0 .$$

Moreover, since $\mathcal{W}(z) \to \mathrm{Id}$ at infinity, and thanks to the special form (2.38) of $\mathcal{V}(\underline{p}, 0)$, $u(z)$ has a finite limit at infinity if and only if $v_2(0) \in \mathbb{G}^P_-(\underline{p}, 0)$ and in this case $\lim u(z) = u_2(0)$.

Therefore, $T_{\underline{u}} \mathcal{C}_{\underline{b}}$ is the set of $u \in \mathbb{C}^N$ such that there is $v \in \mathbb{G}^P_-(\underline{p}, 0)$ such that $U_2(0) = (u, v) \in \ker \Gamma_2(\underline{p}, 0)$. Thus, if $u \in T_{\underline{u}} \mathcal{C}_{\underline{b}} \cap \mathbb{H}_-(\underline{p}, \check\zeta, 0)$, with (1.29), one must have $(u, v) \in \mathbb{G}^0_-(\underline{p}, \check\zeta) \cap \ker \Gamma_2(\underline{p}, 0)$ and therefore $(u, v) = 0$ by Proposition 1.3. This shows that $T_{\underline{u}} \mathcal{C}_{\underline{b}} \cap \mathbb{H}_-(\underline{p}, \check\zeta, 0) = \{0\}$.

By Lemma 2.9, $P(\underline{p}, 0) = (B^\infty_{d,d})^{-1} A^\infty_d$. It has N_- eigenvalues in $\{\mathrm{Re}\, \mu < 0\}$. Thus $\mathbb{G}^P_-(\underline{p}, 0)$ has dimension N_-. Since the total dimension of $\mathbb{G}^0_-(\underline{p}, \check\zeta)$ is N, by (1.29) we deduce that $\mathbb{H}_-(\underline{p}, \check\zeta, 0)$ has dimension $N - N_-$.

Therefore (1.33) holds and the proposition is proved. ☐

1.3. Symmetrizers. Proof of Lemma 2.13. We now proceed to the construction of the symmetrizers. We work in a small neighborhood of $\underline{p}$. We use the block structure of $\mathcal{G}_2$ and construct a symmetrizer for each block separately. We first construct a symmetrizer $S_2(p,\zeta)$ for $P(p,\zeta)$ for p close to $\underline{p}$ and ζ small. Next we construct $\check{S}_1(p,\check{\zeta},\rho)$ adapted to $\check{H}(p,\check{\zeta},\zeta)$, for p close to $\underline{p}$, $|\check{\zeta}|=1$ with $\check{\gamma}\geq 0$ and $\rho\geq 0$. We first argue locally around a given point $\underline{\check{\zeta}}$ of length one, with $\check{\gamma}\geq 0$.

For the estimate $\mathrm{S}+C\Gamma_2^*\Gamma_2\geq c\mathrm{Id}$, for some $c>0$, to hold on a small neighborhood of $(\underline{p},\underline{\check{\zeta}},0)$ it is sufficient that it holds at $(\underline{p},\underline{\check{\zeta}},0)$, since the symbols $\mathrm{S}(p,\check{\zeta},\rho)$ will be C^∞ up to $\rho=0$ and $\check{\gamma}=0$. Thus, by estimate (1.32) of Proposition 1.3, it is sufficient that for all $\kappa_0>0$ large, one can choose $\check{S}_1$ and S_2 such that there is $C>0$, possibly depending on κ_0, such that

$$\big(\mathrm{S}(\underline{p},\underline{\check{\zeta}},0)\big)U,U\big)\geq C\Big(\kappa_0|\Pi_+(\underline{p},\underline{\check{\zeta}})U|^2-|\Pi_-(\underline{p},\underline{\check{\zeta}})U|^2\Big).$$

This is satisfied, if for all κ_0 one can choose $\check{S}_1$ and S_2 and $C>0$ such that

$$\big(\check{S}_1(\underline{p},\underline{\check{\zeta}},0)u,u\big)\geq C\big(\kappa_0|\Pi_+^H(\underline{p},\underline{\check{\zeta}})u|^2-|\Pi_-^H(\underline{p},\underline{\check{\zeta}})u|^2\big),\tag{1.34}$$

$$\big(S_2(\underline{p},0)v,v\big)\geq C\Big(\kappa_0|\Pi_+^P(\underline{p})v|^2-|\Pi_-^P(\underline{p})v|^2\Big).\tag{1.35}$$

a) S_2 is constructed as in the proof of Lemma 2.12. We can assume that

$$P=\begin{bmatrix} P_+ & 0\\ 0 & P_-\end{bmatrix}$$

with $P_\pm$ having their spectrum in $\{\pm\mathrm{Re}\,\mu>0\}$. We choose

$$S_2=\begin{bmatrix} \kappa S_{2,+} & 0\\ 0 & -S_{2,-}\end{bmatrix}$$

with $S_{2,\pm}=S_{2,\pm}^*\geq\mathrm{Id}$ such that $\pm\mathrm{Re}\,(S_{2,\pm}P_\pm)\geq\mathrm{Id}$. If κ is large enough, then (1.35) is satisfied.

b) To construct $\check{S}_1$, we argue similarly as in [**Kr**] [**Ch-P**]. We construct $\check{S}_1$ in the block decomposition of $V^{-1}\check{H}V=\mathrm{Diag}(\mathcal{Q}_k)$ of Lemma 2.10:

$$\check{S}_1=(V^{-1})^*\begin{pmatrix}\mathcal{S}_1 & & \\ & \ddots & \\ & & \mathcal{S}_{\underline{k}}\end{pmatrix}V^{-1}.$$

• If $\mathcal{Q}_k$ satisfies condition $i)$ of Lemma 2.10, we choose $\mathcal{S}_k=\kappa\mathcal{S}_{k,1}$ where $\mathcal{S}_{k,1}=\mathcal{S}_{k,1}^*\geq\mathrm{Id}$ and $\mathrm{Re}\,(\mathcal{S}_{k,1}\mathcal{Q}_k)\geq\mathrm{Id}$.

• If $\mathcal{Q}_k$ satisfies condition $i)$ of Lemma 2.10, we choose $\mathcal{S}_k=\mathcal{S}_k=\mathcal{S}_{k,1}^*\geq\mathrm{Id}$ and $-\mathrm{Re}\,(\mathcal{S}_k\mathcal{Q}_k)\geq\mathrm{Id}$.

• Suppose now that $\underline{\check{\gamma}}=0$ and $\mathcal{Q}_k$ satisfies condition $iii)$. Then $\mathcal{Q}_k(p,\check{\zeta},\rho)=Q_k(p,\check{\zeta})+\rho\mathcal{R}_k(p,\check{\zeta},\rho)$ with Q_k scalar, and purely imaginary when $\check{\gamma}=0$. Thus, $\mathrm{Re}\,Q_k(p,\check{\zeta})=\check{\gamma}Q_k'(p,\check{\zeta})$ with Q_k' smooth near $(\underline{p},\underline{\check{\zeta}})$.

1) If $\partial_\gamma Q_k(\underline{p},\underline{\check{\zeta}})<0$ and $\mathcal{R}(\underline{p},\underline{\check{\zeta}},0)$ is negative definite, we choose $\mathcal{S}_k(p,\check{\zeta},\rho)=-\mathrm{Id}$ so that

$$\mathrm{Re}\,(-\mathcal{Q}_k)=\check{\gamma}(-Q_k'\mathrm{Id})+\rho\mathrm{Re}\,(-\mathcal{R}_k)$$

where $(-Q_k'\mathrm{Id}$ and $\mathrm{Re}\,(-\mathcal{R}_k)$ are positive definite at $(\underline{p},\underline{\check{\zeta}},0)$.

2) If $\partial_\gamma Q_k(\underline{p},\underline{\check\zeta}) > 0$ and $\mathcal{R}(\underline{p},\underline{\check\zeta},0)$ is positive definite, we choose $\mathcal{S}_k(p,\check\zeta,\rho) = \kappa\mathrm{Id}$ so that

$$\mathrm{Re}\,(\mathcal{Q}_k) = \kappa\check\gamma(Q'_k\mathrm{Id}) + \rho\mathrm{Re}\,(\mathcal{R}_k)$$

where $(Q'_k\mathrm{Id}$ and $\mathrm{Re}\,(\mathcal{R}_k)$ are positive definite at $(\underline{p},\underline{\check\zeta},0)$.

• We now come to the most delicate part when $\underline{\check\gamma} = 0$ and $\mathcal{Q}_k$ satisfies condition *iv*). Next, in the block reduction of $\mathcal{Q}_k$, we choose the $\mathcal{S}_k$ diagonal:

$$\mathcal{S}_k = \begin{bmatrix} S_k & 0 & \\ 0 & S_k & \\ & & \ddots \end{bmatrix}, \tag{1.36}$$

$$S_k(p,\check\zeta,\rho) = E_k + \widetilde{E}_k(p,\check\zeta) - i\gamma F_k - i\rho F'_k$$

where E_k and $\widetilde{E}_k$ are real symmetric matrices, and F_k and H_k are real and skew symmetric. Moreover, E_k have the special form

$$E_k = \begin{bmatrix} 0 & \cdots & \cdots & 0 & e_{k,1} \\ \vdots & & & \cdot^{\cdot^{\cdot}} & e_{k,2} \\ \vdots & & \cdot^{\cdot^{\cdot}} & \cdot^{\cdot^{\cdot}} & \\ 0 & \cdot^{\cdot^{\cdot}} & \cdot^{\cdot^{\cdot}} & & \\ e_{k,1} & e_{k,2} & & & e_{k,\nu_k} \end{bmatrix},$$

and $\widetilde{E}_k(\underline{p},\underline{\check\zeta}) = 0$.

The order of the construction is as follows. One first chooses E_k, $\widetilde{E}_k$ and F_k as in [**Kr**] to construct a symmetrizer for Q_k, that is for $\rho = 0$. The new part lies in the choice of F'_k.

1. Choose E_k such that

$$\mathrm{Re}\,\big(E_k\partial_\gamma Q_k(\underline{p},\underline{\check\zeta})w, w\big) \geq 2|w_1|^2 - C|w'|^2 \tag{1.37}$$

with w_1 the first component of $w \in \mathbb{C}^{\nu_k}$ and $w' \in \mathbb{C}^{\nu_k-1}$ denotes the other components, and

$$(E_kU, U) \geq C_1\Big(\kappa|\Pi^k_+U|^2 - |\Pi^k_-U|^2\Big), \tag{1.38}$$

where $\Pi^k_\pm$ is the projection onto $\mathrm{H}^k_\pm(\underline{p},\underline{\check\zeta})$ in the decomposition $\mathbb{C}^{\nu_k} = \mathrm{H}^k_+ \oplus \mathrm{H}^k_-$. This is one of the basic points of the construction in [**Kr**]. According to [**Ch-P**], one first chooses the coefficient $e_{k,1}$ such that

$$e_{k,1}\partial_\gamma\mathrm{Re}\,a_k(\underline{p},\underline{\check\zeta}) \geq 3. \tag{1.39}$$

This is sufficient to imply (1.37) (cf (equation (5.5.3) in [**Ch-P**], Chap. 7). Next, the coefficients $e_{k,l}$ for $l > 1$ are chosen successively to achieve (1.38) (cf Lemma 5.6 in [**Ch-P**], Chap. 7).[2] Note that the constant C in (1.37) depends the coefficients $e_{k,l}$ thus on κ, but the condition (1.39) is independent of κ.

2. Recall that $Q_k(\underline{p},\underline{\check\zeta}) = i(\mu_k\mathrm{Id} + N_k)$ with $\mu_k \in \mathbb{R}$ and N_k the Jordan's matrix of size ν_k. The form of E_k is chosen so that $E_k(\mu_k\mathrm{Id} + N_k)$ is real and symmetric. Next, the real matrix $\widetilde{E}_k(p,\check\zeta)$ is chosen so that such that $(E_k + \widetilde{E}_k)(\frac{1}{i}Q_k)$ is

[2]The reader must be aware that the symbol of our symmetrizer is the opposite of the symbol constructed in [**Ch-P**], where the symbol of the equation A is $\frac{1}{i}\check H$ in our notation, so that $\mathrm{Re}\,(SH) = \mathrm{Im}\,(-SA)$.

real and symmetric when $\check{\gamma} = 0$. This is achieved in [**Kr**] [**Ch-P**] using the implicit function theorem and the property that $\frac{1}{i}Q_k$ is real when $\check{\gamma} = 0$.

3. Following [**Kr**] [**Ch-P**], for all C, there is F_k real and skew symmetric such that

$$\mathrm{Re}\,(F_k N_k w, w) \geq -|w_1|^2 + (C+1)|w'|^2.$$

We take C the constant found in (1.37). As a consequence, we have

$$\begin{aligned} &\mathrm{Re}\,\big((E_k + \widetilde{E}_k - i\check{\gamma}F_k)Q_k\big) = \check{\gamma}D_k\,, \\ &D_k(\underline{p}, \underline{\check{\zeta}}) = \mathrm{Re}\,(E_k\partial_\gamma Q_k(\underline{p}, \underline{\check{\zeta}}) + \mathrm{Re}\,(F_kN_k) \geq \mathrm{Id}\,. \end{aligned} \tag{1.40}$$

4. We come to the new part. Denote by $\mathcal{E}_k$ the block diagonal matrix $\mathrm{Diag}(E_k)$. A vector $w \in \mathbb{C}^{N_k}$, $N_k = \nu_k\alpha_k$ being the dimension of $\mathcal{Q}_k$, is broken into α_k blocks $w_p \in \mathbb{C}^{\nu_k}$, with components denoted by $w_{p,a}$. We denote by $R_{p,q}$ the $\nu_k \times \nu_k$ blocks of $\mathcal{R}_k$ and by $R_{p,a,q,b}$ their entries. The entries of E_k are denoted by $E_{a,b}$. Then, since $R_{p,a,q,b} = 0$ when $b > 1$ and taking into account the special form of E_k,

$$\begin{aligned} \mathrm{Re}\,(\mathcal{E}_k\mathcal{R}_k w, w) &= \mathrm{Re} \sum E_{a,c}R_{p,a,q,1}w_{q,1}\overline{w}_{p,c} \\ &= \mathrm{Re} \sum e_{k,1}r_{p,q}w_{q,1}\overline{w}_{p,1} + O(|w_{*,1}|\,|w'_*|) \end{aligned}$$

where $w_{*,1} \in \mathbb{C}^{\alpha_k}$ is the collection of the first components $w_{p,1}$, w'_* the remainder components and $r_{p,q} = R_{p,\nu_k,q,1}$ the lower left hand corner entry of $R_{p,q}$. The matrix $\mathrm{Re}\,(R^\flat_k)$ is definite, positive or negative according to the sign of $\partial_\gamma a_k$. Moreover $e_{k,1}$ has the sign of $\partial_\gamma a_k$ by (1.39). Thus, multiplying E_k by some positive constant, we can achieve that in addition to (1.38) (1.40), the following inequality holds:

$$\mathrm{Re}\,\big(\mathcal{E}_k\mathcal{R}_k(\underline{p}, \underline{\check{\zeta}}, 0)w, w\big) \geq 2|w_{*,1}|^2 - C'|w'_*|^2. \tag{1.41}$$

5. Next, as in 3, there is F'_k real and skew symmetric such that for all $w \in \mathbb{C}^{\nu_k}$:

$$\mathrm{Re}\,(F'_kN_kw, w) \geq -|w_1|^2 + (C'+1)|w'|^2.$$

Thus, with $\mathcal{F}'_k = \mathrm{Diag}(F'_k)$, $\mathcal{N}_k = \mathrm{Diag}(N_k)$ and $w \in \mathbb{C}^{N_k}$:

$$\mathrm{Re}\,(\mathcal{F}'_k\mathcal{N}_kw, w) \geq -|w_{*,1}|^2 + (C'+1)|w'_*|^2.$$

Therefore, with (1.41), we have

$$\mathrm{Re}\,\big(\mathcal{E}_k\mathcal{R}_k(\underline{p}, \underline{\check{\zeta}}, 0) - i\mathcal{F}'_k\mathcal{Q}_k(\underline{p}, \underline{\check{\zeta}})\big) \geq \mathrm{Id}\,. \tag{1.42}$$

Adding up, we see that the matrix defined in (1.36) satisfies

$$\mathrm{Re}\,\big(\mathcal{S}_k(\mathcal{Q}_k + \rho\mathcal{R}_k)\big) = \check{\gamma}\mathcal{D}_k(p, \check{\zeta}) + \rho\mathcal{D}'_k(p, \check{\zeta}, \rho)$$

with $\mathcal{D}_k = \mathrm{Diag}(D_k)$ and at the base point

$$\mathcal{D}'_k(\underline{p}, \underline{\check{\zeta}}, 0) = \mathrm{Re}\,\big(\mathcal{E}_k\mathcal{R}_k(\underline{p}, \underline{\check{\zeta}}, 0) - i\mathcal{F}'_k\mathcal{Q}_k(\underline{p}, \underline{\check{\zeta}})\big)\,.$$

By (1.40) and (1.42), the matrices $\mathcal{D}_k$ and $\mathcal{D}'_k$ are positive definite on a neighborhood of the base point $(\underline{p}, \underline{\check{\zeta}}, 0)$.

This shows that the symmetric matrix $\check{S}_1 = (V^{-1})\mathrm{Diag}(\mathcal{S}_k)V^{-1}$ is defined on a neighborhood of the base point $(\underline{p}, \underline{\check{\zeta}}, 0)$ and satisfies

$$\mathrm{Re}\,(\check{S}_1\check{H}) = (V^{-1})\mathrm{Diag}\big(\mathrm{Re}\,\mathcal{S}_k(\mathcal{Q}_k + \rho\mathcal{R}_k)\big)V^{-1}$$

and the blocks $\mathrm{Re}\,\mathcal{S}_k(\mathcal{Q}_k+\rho\mathcal{R}_k)$ are either positive definite or of the form $\check{\gamma}\mathcal{D}_k+\rho\mathcal{D}'_k$ with $\mathcal{D}_k$ and $\mathcal{D}'_k$ are positive definite. Moreover, recalling the form of the spaces $\mathbb{H}_-$ and $\mathbb{H}_+$, we see that the condition (1.34) is satisfied on a neighborhood $(\underline{p},\underline{\check{\zeta}},0)$.

c) So far we have constructed $\mathcal{S}_1$ on a neighborhood of a given point $(\underline{p},\underline{\check{\zeta}},0)$ with $\underline{\check{\zeta}}$ in the closed half unit sphere $S^{d+1}_+ = \{|\check{\zeta}|=1\,,\check{\gamma}\geq 0\}$. Using a partition of unity on S^{d+1}_+, we define $\check{S}_1 = \sum\phi_l(V_l^{-1})^*\widetilde{S}_lV_l^{-1}\phi_l$ where ϕ_l is nonnegative, smooth and supported on a small neighborhood of $(\underline{p},\underline{\check{\zeta}}_l,0)$, $\widetilde{H}_l := V_l^{-1}\check{H}V_l$ block diagonal on this neighborhood and $\widetilde{S}_l$ is block diagonal as in part b) above. Moreover, $\sum\phi_l^2 = 1$ on a neighborhood of $\{\underline{p}\}\times S^{d+1}_+\times\{0\}$. We see that $\sum\phi_l^2(V_l^{-1})^*V_l^{-1}$ is positive definite on this neighborhood, that $\mathrm{Re}\,(\mathcal{S}_1\check{H}) = \sum\phi_l(V_l^{-1})^*\mathrm{Re}\,(\widetilde{S}_l\widetilde{H}_l)V_l^{-1}\phi_l$ satisfies the properties listed in part *ii)* of Lemma 2.13. In addition, because the ϕ_l are non negative and $\sum\phi_l^2=1$, the condition (1.34) holds on a neighborhood of $\{\underline{p}\}\times S^{d+1}_+\times\{0\}$.

The proof of Lemma 2.13 is now complete.

2. Appendix B. Para-differential calculus

In this appendix, we prove the different results stated in section 3. Most of the analysis follows known results from [**Bo**] [**Mey**] (see also [**Hör**] [**Tay**]) for the classical calculus and from [**Mok**] [**Mé1**] for the calculus with a large parameter γ. Most of the present work is to check that these results extend to symbols with parabolic homogeneity and next to a semi-classical calculus.

2.1. The quasi-homogeneous calculus. To include in the same analysis both the homogeneous case and the parabolic case, we first consider a general quasi-homogeneous framework. With little risk of confusion with our previous notation, we denote by x the variable in $\mathbb{R}^d$, and ξ its dual variable.

For given positive integers p_j, consider on $\mathbb{R}^{1+d}$ the quasi-homogeneous pseudo-norm

$$\text{for } \zeta=(\xi,\gamma)\ :\quad \langle\zeta\rangle = \Big(\gamma^{2p_0}+\sum_{j=1}^d \xi_j^{2p_j}\Big)^{1/2p}\,,\qquad p:=\max p_j. \tag{2.1}$$

The case $p_j=p=1$ for all j corresponds to the usual Euclidean norm. It will be referred to as the homogeneous case. The parabolic case corresponds to $p_j=2$ for the spatial directions and $p_j=1$ for the time direction. We agree that $\langle\xi\rangle=\langle\xi,0\rangle$. We also introduce the weight

$$\Lambda(\zeta) = (1+\langle\zeta\rangle^{2p})^{1/2p}. \tag{2.2}$$

Note that

$$\langle\zeta+\zeta'\rangle\leq\langle\zeta\rangle+\langle\zeta'\rangle\,,\quad \Lambda(\zeta+\zeta')\leq\Lambda(\zeta)+\langle\zeta'\rangle\,. \tag{2.3}$$

Introduce the quasi-homogeneous dilations

$$\rho\cdot(\xi,\gamma) = (\rho^{p/p_1}\xi_1,\ldots,\rho^{p/p_d}\xi_d,\rho^{p/p_0}\gamma) \tag{2.4}$$

and define similarly $\rho\cdot\xi$. Then

$$\langle\rho\cdot\zeta\rangle = \rho\langle\zeta\rangle\,. \tag{2.5}$$

The pseudo-norm defines a distance $d(\xi,\eta) = \langle \xi - \eta \rangle$. Denote by $B(\eta,\rho)$ the ball centered at η of radius ρ, i.e. the set of $\eta' \in \mathbb{R}^d$ such that $\langle \eta - \eta' \rangle \le \rho$. The following properties are elementary:

$$\text{(2.6)} \qquad \mathrm{meas} B(\eta,\rho) = \rho^D \mathrm{meas} B(\eta,1) \quad \text{with } D = \sum_1^d p/p_j \, .$$

(2.7) There is an integer N, such that for all η and ρ, the ball $B(\eta,\rho)$ is contained in the union of at most N balls of radius $\rho/2$.

The dilations and quasi-homogeneous norms are also defined in the x space :

$$\rho\cdot x = (\rho^{p/p_1} x_1, \dots, \rho^{p/p_d} x_d)\,, \qquad \langle x \rangle = \Big(\sum_{j=1}^d (x_j)^{2p_j} \Big)^{1/2p} .$$

2.1.1. *The Littlewood-Paley decomposition.* Introduce $\chi \in C_0^\infty(\mathbb{R})$, such that $0 \le \chi \le 1$ and

$$\text{(2.8)} \qquad \chi(\lambda) = 1 \quad \text{for } |\lambda| \le 1.1\,, \qquad \chi(\lambda) = 0 \quad \text{for } |\lambda| \ge 1.9\,.$$

For $k \in \mathbb{Z}$, introduce $\chi_k(\xi,\gamma) := \chi(2^{-k}\Lambda(\xi,\gamma))$, $\widetilde{\chi}_k^\gamma(x)$ its inverse Fourier transform with respect to ξ and the operators

$$\text{(2.9)} \qquad S_k^\gamma u := \widetilde{\chi}_k^\gamma * u = \chi_k(D_x,\gamma)u\,, \quad \Delta_{k+1}^\gamma = S_{k+1}^\gamma - S_k^\gamma .$$

For all temperate distribution u, the spectrum of $\Delta_k^\gamma u$ (i.e. the support of its Fourier transform) satisfies

$$\text{(2.10)} \qquad \mathrm{spec}(\Delta_k^\gamma u) \subset \big\{ \xi \ : \ 2^{k-1} < \Lambda(\xi,\gamma) < 2^{k+1} \big\} .$$

Hence $\Delta_k^\gamma u = 0$ when $(1+\gamma^{2p_0})^{1/2p} \ge 2^{k+1}$, and in particular when $k < 0$. Thus, for all temperate distribution u, one has for $\gamma \ge 0$:

$$\text{(2.11)} \qquad u = \sum_{k\ge 0} \Delta_k^\gamma u\, .$$

The natural Sololev spaces associated to the weights Λ^s, $s \in \mathbb{R}$, are the spaces $\mathrm{H}^s(\mathbb{R}^d)$ of temperate distributions u such that their Fourier transform $\widehat{u}$ satisfies $\Lambda(\xi,0)^s \widehat{u} \in L^2(\mathbb{R}^d)$. This space is equipped with the family of norms :

$$\text{(2.12)} \qquad \|u\|_{s,\gamma}^2 := \int \Lambda^{2s}(\xi,\gamma) |\widehat{u}(\xi)|^2 \, d\xi\, .$$

The next propositions immediately follow from the definitions. The important point is that the constants C do not depend on $\gamma \ge 0$.

PROPOSITION 2.1. *Consider $s \in \mathbb{R}$ and $\gamma \ge 0$. A temperate distribution u belongs to $\mathrm{H}^s(\mathbb{R}^d)$ if and only if*

i) for all $k \in \mathbb{N}$, $\Delta_k^\gamma u \in L^2(\mathbb{R}^d)$.

ii) the sequence $\delta_k = 2^{ks}\|\Delta_k^\gamma u\|_{L^2(\mathbb{R}^d)}$ belongs to $\ell^2(\mathbb{N})$.

Moreover, there is a constant C, independent of $\gamma \ge 0$ and u in H^s, such that

$$\frac{1}{C}\|u\|_{s,\gamma}^2 \le \sum_{k\ge 0} \delta_k^2 \le C\|u\|_{s,\gamma}^2$$

PROPOSITION 2.2. *Consider $s \in \mathbb{R}$, $\gamma \geq 0$ and $R > 0$. Suppose that $\{u_k\}_{k\in\mathbb{N}}$ is a sequence of functions in $L^2(\mathbb{R}^d)$ such that:*

i) the spectrum of u_k is contained in $\{\frac{1}{R}2^k \leq \Lambda(\xi,\gamma) \leq R2^k\}$.

ii) the sequence $\delta_k = 2^{ks}\|u_k\|_{L^2(\mathbb{R}^d)}$ belongs to $\ell^2(\mathbb{N})$.

Then $u = \sum u_k$ belongs to $\mathrm{H}^s(\mathbb{R}^d)$ and there is a constant C, independent of $\gamma \geq 0$ and the sequence $\{u_k\}$, such that

$$\|u\|_{s,\gamma}^2 \leq C \sum_k \delta_k^2 .$$

When $s > 0$, it is sufficient to assume that the spectrum of u_k is contained in $\{\Lambda(\xi,\gamma) \leq R2^k\}$.

We also use the space $W^{1,\infty}(\mathbb{R}^d)$ of functions $u \in L^\infty$ such that $\nabla_x u \in L^\infty$. It is equipped with the obvious norm. With χ as in (2.8), we denote by $\dot S_k$, $\dot\Delta_k$ the Littlewood-Paley decomposition associated to the Fourier multipliers $\dot\chi_k(\xi) = \chi(2^{-k}\langle\xi\rangle) = \chi(\langle 2^{-k}\cdot\xi\rangle)$, that is with $\gamma = 0$:

$$\dot S_k = \dot\chi_k(D_x), \quad \dot\Delta_k = \dot S_k - \dot S_{k-1} . \tag{2.13}$$

PROPOSITION 2.3. *There is a constant C such that :*

i) for all $u \in L^\infty$ and all $k \in \mathbb{Z}$, one has

$$\|\dot S_k u\|_{L^\infty} \leq C\|u\|_{L^\infty} .$$

ii) for all $u \in W^{1,\infty}$ and all $k \in \mathbb{N}$, one has

$$\|\dot\Delta_k u\|_{L^\infty} \leq C2^{-k}\|\nabla u\|_{L^\infty} , \quad \|u - \dot S_k u\|_{L^\infty} \leq C2^{-k}\|\nabla_x u\|_{L^\infty} .$$

PROOF. $\dot S_k$ is a convolution operator with kernel $\dot S_k(x) = 2^{kD}\dot S_0(2^k\cdot x)$ where $\dot S_0(\cdot)$ is the inverse Fourier transform of $\dot\chi_0(\xi) = \chi(\langle\xi\rangle)$. Since $\dot\chi_0 \in C_0^\infty(\mathbb{R}^d)$, $\dot S_0(\cdot)$ belongs to the Schwartz class, hence to L^1. Thus $\dot S_k(\cdot) \in L^1$ and $\|\dot S_k\|_{L^1} = \|\dot S_0\|_{L^1}$, implying i).

Since $\chi(0) = 1$, the integral of the kernel $\dot S_k$ is one and

$$\begin{aligned} |u(x) - \dot S_k u(x)| &= \Big|\int \dot S_k(y)(u(x) - u(x-y))dy\Big| \\ &\leq \|\nabla u\|_{L^\infty} \int |\dot S_k(y)|\,|y|\,dy \end{aligned}$$

where $|y|$ denotes the Euclidean norm of $y \in \mathbb{R}^n$. Then the second estimate in ii) follows from the inequalities

$$\begin{aligned} \int |\dot S_k(y)|\,|y|\,dy &= \int |\dot S_0(y)||2^{-k}\cdot y|dy \\ &\leq \sum_j 2^{-kp/p_j} \int |\dot S_0(y)|\,|y_j|dy \leq C2^{-k}. \end{aligned}$$

The proof of the estimate for $\dot\Delta_k u$ is similar. □

2.1.2. *Paradifferential operators with parameters.* If $a(\zeta)$ is smooth then

$$\partial_\xi^\alpha(a(\lambda\cdot\zeta)) = \lambda^{\langle\alpha\rangle}(\partial_\xi^\alpha a)(\lambda\cdot x)$$

where

$$\text{(2.14)} \qquad \text{for} \quad \alpha \in \mathbb{N}^d : \quad \langle \alpha \rangle = \sum_{j=1}^{d} \frac{p}{p_j} \alpha_j \, .$$

In particular, if $a(\zeta)$ is smooth and quasi-homogeneous of degree μ for $\zeta \neq 0$, meaning that $a(\rho \cdot \zeta) = \rho^\mu a(\zeta)$ for all $\rho > 0$, then its derivatives $\partial_\xi^\alpha a$ are quasi-homogeneous of degree $\mu - \langle \alpha \rangle$ and thus are bounded by $C \langle \zeta \rangle^{\mu - \langle \alpha \rangle}$. Similarly, for all $\alpha \in \mathbb{N}^d$, there is C_α such that

$$\text{(2.15)} \qquad \forall (\xi, \gamma) : \quad |\partial_\xi^\alpha \Lambda(\xi, \gamma)| \leq C_\alpha \Lambda(\xi, \gamma)^{1 - \langle \alpha \rangle} \, .$$

This motivates the following definition.

DEFINITION 2.4 (Symbols). Let $\mu \in \mathbb{R}$.

i) Γ_0^μ denotes the space of locally bounded functions $a(x, \xi, \gamma)$ on $\mathbb{R}^d \times \mathbb{R}^d \times [0, \infty[$ which are C^∞ with respect to ξ and such that for all $\alpha \in \mathbb{N}^d$ there is a constant C_α such that

$$\text{(2.16)} \qquad \forall (x, \xi, \gamma) , \quad |\partial_\xi^\alpha a(x, \xi, \gamma)| \leq C_\alpha \, \Lambda(\xi, \gamma)^{\mu - \langle \alpha \rangle} .$$

ii) Γ_1^μ denotes the space of symbols $a \in \Gamma_0^\mu$ such that for all j, $\partial_{x_j} a \in \Gamma_0^\mu$.

iii) For $k = 0, 1$, Σ_k^μ is the space of symbols $\sigma \in \Gamma_k^\mu$ such that there exists $\delta \in]0, 1[$ such that for all (ξ, γ) the spectrum of $x \mapsto \sigma(x, \xi, \gamma)$ is contained in the quasi-homogeneous ball $\{ \langle \eta \rangle \leq \delta \Lambda(\xi, \gamma) \}$.

The spaces Γ_0^μ are equipped with semi-norms

$$\text{(2.17)} \qquad \|a\|_{(\mu, N)} := \sup_{|\alpha| \leq N} \; \sup_{\mathbb{R}^d \times \mathbb{R}^d \times [0, \infty[} \; \Lambda(\xi, \gamma)^{\langle \alpha \rangle - \mu} |\partial_\xi^\alpha a(x, \xi, \gamma)| \, .$$

Consider a C^∞ function $\psi(\eta, \xi, \gamma)$ on $\mathbb{R}^d \times \mathbb{R}^d \times [0, \infty[$ such that:

1) there are δ_1 and δ_2 such that $0 < \delta_1 < \delta_2 < 1$ and

$$\text{(2.18)} \qquad \begin{cases} \psi(\eta, \xi, \gamma) = 1 & \text{for } \langle \eta \rangle \leq \delta_1 \Lambda(\xi, \gamma) \\ \psi(\eta, \xi, \gamma) = 0 & \text{for } \langle \eta \rangle \geq \delta_2 \Lambda(\xi, \gamma) \, . \end{cases}$$

2) for all $(\alpha, \beta) \in \mathbb{N}^d \times \mathbb{N}^d$, there is $C_{\alpha, \beta}$ such that

$$\text{(2.19)} \qquad \forall (\eta, \xi, \gamma) , \quad |\partial_\xi^\alpha \partial_\eta^\beta \psi(\eta, \xi, \gamma)| \leq C_{\alpha, \beta} \Lambda(\xi, \gamma)^{-\langle \alpha \rangle - \langle \beta \rangle} \, .$$

For instance, with $N \geq 3$, $\delta_1 = 2^{-N-2}$ and $\delta_2 = 2^{2-N}$, one can consider

$$\text{(2.20)} \qquad \psi(\eta, \xi, \gamma) = \sum_{k \geq 0} \chi(2^{-k+N} \langle \eta \rangle)(\chi_k(\xi, \gamma) - \chi_{k-1}(\xi, \gamma)).$$

We will say that such a function ψ is an *admissible cut-off*. Consider next $G^\psi(\,\cdot\,, \xi, \gamma)$ the inverse Fourier transform of $\psi(\,\cdot\,, \xi, \gamma)$. It satisfies

$$\text{(2.21)} \qquad \forall (\xi, \gamma) , \quad \|\partial_\xi^\alpha G^\psi(\,\cdot\,, \xi, \gamma)\|_{L^1(\mathbb{R}^d)} \leq C_\alpha \Lambda(\xi, \gamma)^{-\langle \alpha \rangle} \, .$$

Indeed, $G^\psi(x, \zeta) = \Lambda^D G_\zeta^\flat(\Lambda \cdot x)$ where $G_\zeta^\flat$ is the inverse Fourier transform of $\psi_\zeta^\flat(\eta) = \psi(\Lambda \cdot \eta, \zeta)$ and $\Lambda = \Lambda(\zeta)$. The estimates (2.19) imply that the $\psi_\zeta^\flat$ are bounded in $C^\infty(\mathbb{R}^d)$ with support in the ball $\{ \langle \eta \rangle \leq \delta_2 \}$. Thus the $G_\zeta^\flat$ and hence the $G^\psi(\cdot, \zeta)$ are uniformly bounded in L^1. The analysis of the ξ derivatives is analogous.

The argument above applied to a fixed ζ, implies the following version of Bernstein's inequality which is used at several places:

LEMMA 2.5. *If the spectrum of $a \in L^p(\mathbb{R}^d)$ is contained in the ball $\{\langle\xi\rangle \leq \lambda\}$, then $a \in C^\infty$ and for all $\alpha \in \mathbb{N}^d$*

$$\|\partial_x^\alpha a\|_{L^p} \leq C\lambda^{\langle\alpha\rangle}\|a\|_{L^p}$$

with C_α independent of a and λ.

REMARK 2.6. The main difference with [**Bo**] [**Mey**] in the choice of admissible cut-off functions ψ (2.18) is the treatment of low frequencies: we assume that ψ is one for small η, even when ξ is small. This has no importance in the study of smoothness, but it plays a crucial role in the semi-classical version of the calculus and also when one considers parameters to absorb lower order terms.

PROPOSITION 2.7. *Let ψ be an admissible cut-off. Then, for all $\mu \in \mathbb{R}$ and $k = 0, 1$, the operators*

$$a \mapsto \sigma_a^\psi(x,\xi,\gamma) := \int G^\psi(x-y,\xi,\gamma)\, a(y,\xi,\gamma)\, dy \tag{2.22}$$

are bounded from Γ_k^μ to Σ_k^μ and

$$\|\sigma_a^\psi\|_{(\mu,N)} \leq C_N\|a\|_{(\mu,N)}\,.$$

Moreover, if $a \in \Gamma_1^\mu$, then $a - \sigma_a^\psi \in \Gamma_0^{\mu-1}$. In particular, if ψ_1 and ψ_2 are admissible and $a \in \Gamma_1^\mu$ then $\sigma_a^{\psi_1} - \sigma_a^{\psi_2} \in \Sigma_0^{\mu-1}$. More precisely

$$\|\sigma_a^{\psi_1} - \sigma_a^{\psi_2}\|_{(\mu-1,N)} \leq C_N\|\nabla_x a\|_{(\mu,N)}\,.$$

PROOF. The bounds (2.21) imply that the estimates (2.16) are preserved by the convolution (2.22). Thus $\sigma_a^\psi \in \Gamma_0^\mu$ if $a \in \Gamma_0^\mu$. Moreover, $\partial_x\sigma_a^\psi = \sigma_{\partial_x a}^\psi$ and the operator (2.22) maps Γ_1^μ into itself. On the Fourier side, one has

$$\widehat{\sigma}_a^\psi(\eta,\xi,\gamma) = \psi(\eta,\xi,\gamma)\,\widehat{a}(\eta,\xi,\gamma).$$

Thus, the spectral property is clear and the first part of the proposition is proved.

For fixed $\zeta = (\xi,\gamma)$, the mapping $a(\cdot,\zeta) \mapsto \sigma_a^\psi(\cdot,\zeta)$ is a convolution operator with the inverse Fourier transform of $\psi(\cdot,\zeta)$. The estimates (2.19) imply that the family of mappings $\eta \mapsto \psi(\Lambda\cdot\eta,\zeta)$ with $\Lambda = \Lambda(\zeta)$, is bounded in $C_0^\infty(\mathbb{R}^d)$ with support in a fixed ball. Therefore, arguing as in the proof of Proposition 2.3 part two, one shows that

$$\|(a-\sigma_a^\psi)(\,\cdot\,,\xi,\gamma)\|_{L^\infty} \leq C\,\Lambda(\xi,\gamma)^{-1}\|\nabla_x a(\,\cdot\,,\xi,\gamma)\|_{L^\infty}.$$

One has similar estimates for the ξ-derivatives. Thus $a - \sigma_a^\psi \in \Gamma_0^{\mu-1}$ if $a \in \Gamma_1^\mu$. □

The spectral property and Lemma 2.5 imply that the symbols $\sigma \in \Sigma_0^\mu$ are C^∞ in x too and satisfy:

$$|\partial_x^\beta\partial_\xi^\alpha\sigma(x,\xi,\gamma)| \leq C_{\alpha\,\beta}\Lambda^{\mu-\langle\alpha\rangle+\langle\beta\rangle}. \tag{2.23}$$

The associated pseudo-differential operators are defined as

$$\mathrm{Op}^\gamma(\sigma)u(x) := \frac{1}{(2\pi)^d}\int e^{i\xi\cdot x}\,\sigma(x,\xi,\gamma)\,\widehat{u}(\xi)\,d\xi \tag{2.24}$$

or on the Fourier side

$$\widehat{\mathrm{Op}^\gamma(\sigma)u}(\xi) = \int \widehat{\sigma}^1(\xi-\xi',\xi',\gamma)\,\widehat{u}(\xi')\,d\xi' \tag{2.25}$$

where $\widehat{\sigma}^1$ denotes the Fourier transform of σ with respect to x. They act continuously in the Schwartz class. Using Proposition 2.7 we can associate operators to symbols $a \in \Gamma_0^\mu$. Given an admissible cut-off ψ, define

$$T_a^{\psi,\gamma} u := \mathrm{Op}^\gamma(\sigma_a^\psi)\, u. \tag{2.26}$$

Introduce the following terminology.

DEFINITION 2.8. A family of operators $\{P^\gamma\}$ is of order less than or equal to μ if for all $s \in \mathbb{R}$, P^γ maps H^s into $\mathrm{H}^{s-\mu}$ and there is a constant C such that

$$\forall \gamma \geq 0\,,\ \forall u \in \mathrm{H}^s(\mathbb{R}^n) : \quad \|P^\gamma u\|_{s-\mu,\gamma} \leq C\|u\|_{s,\gamma}.$$

PROPOSITION 2.9. *i) For all $\sigma \in \Sigma_0^\mu$, the family of operators $\mathrm{Op}^\gamma(\sigma)$ extends as a family of operators of order $\leq \mu$. More precisely, one has*

$$\|\mathrm{Op}^\gamma(\sigma)u\|_{s,\gamma} \leq C\|\sigma\|_{(\mu,N)}\ \|u\|_{s+\mu,\gamma}$$

where C and N only depend on the indices s and m and on the confinement parameter δ of σ.

ii) For all admissible cut-off ψ and all $a \in \Gamma_0^\mu$, the family of operators $T_a^{\psi,\gamma}$ is of order $\leq \mu$.

iii) If ψ_1 and ψ_2 are admissible and $a \in \Gamma_1^\mu$, then $T_a^{\psi_1,\gamma} - T_a^{\psi_2,\gamma}$ is of order $\leq \mu - 1$. More precisely, one has

$$\|(T_a^{\psi_1,\gamma} - T_a^{\psi_2,\gamma})u\|_{s,\gamma} \leq C\|\nabla_x a\|_{(\mu,N)}\ \|u\|_{s+\mu-1,\gamma}$$

where C and N only depend on the indices s and μ and on the confinement parameters δ of ψ_1 and ψ_2.

PROOF. Using (2.11), one obtains that

$$\mathrm{Op}^\gamma(\sigma) = \sum_{|k-l|\leq 2} \mathrm{Op}^\gamma(\sigma_k)\Delta_l^\gamma$$

where

$$\sigma_k(x,\xi,\gamma) = \big(\chi_k(\xi,\gamma) - \chi_{k-1}(\xi,\gamma)\big)\,\sigma(x,\xi,\gamma)\,.$$

The estimates (2.23) imply that

$$\widetilde{\sigma}_k(x,\xi,\gamma) = \sigma_k\big(2^{-k}\cdot x, 2^k\cdot(\xi,\gamma)\big)$$

is supported in $\{\langle\zeta\rangle \leq 2\}$ and satisfies

$$\big|\partial_x^\alpha \partial_\xi^\beta \widetilde{\sigma}_k(x,\xi,\gamma)\big| \leq C_{\alpha,\beta}\, 2^{k\mu}.$$

Therefore, by Calderon-Vaillancourt's theorem, the operators $\mathrm{Op}^{\gamma'}(\widetilde{\sigma}_k)$ are bounded in L^2 with norm $O(2^{k\mu})$ (see e.g. [**Co-Me**]). Since $\mathrm{Op}^\gamma(\sigma_k) = H_k \mathrm{Op}^{\gamma'}(\widetilde{\sigma}_k)H_k^{-1}$ with $\gamma' = 2^{-kp/p_0}\gamma$ and $H_k u = u(2^k\cdot x)$, this implies that

$$\|\mathrm{Op}^\gamma(\sigma_k)u\|_{L^2} \leq C\,\|\sigma\|_{(\mu,N)} 2^{k\mu}\,\|u\|_{L^2}\,.$$

Moreover, by (2.25), the spectrum of $\mathrm{Op}^\gamma(\sigma_k)u$ is contained in the set of ξ such that there is ξ' satisfying $\langle\xi - \xi'\rangle \leq \delta\Lambda(\xi',\gamma)$ and $2^{k-1} \leq \langle\Lambda(\xi',\gamma)\rangle \leq 2^{k+1}$. Hence, it is contained in the domain

$$\big\{(1-\delta)2^{k-1} \leq \Lambda(\xi,\gamma) \leq (1+\delta)2^{k+1}\big\}\,.$$

Proposition 2.2 implies that $u_k = \sum_{|l-k|\leq 2}\Delta_l^\gamma u$ satisfies

$$\|u_k\|_{L^2} \leq C2^{-ks}\,\varepsilon_k\,\|u\|_{s,\gamma} \quad \text{with} \quad \sum \varepsilon_k^2 \leq 1\,.$$

Therefore

$$\|\mathrm{Op}^\gamma(\sigma_k)\, u_k\|_{L^2} \le C 2^{k(\mu-s)}\, \varepsilon_k\, \|\sigma\|_{(\mu,N)}\, \|u\|_{s,\gamma}\,.$$

Using Proposition 2.3, these estimates and the spectral localization imply i). The other two parts follow from Proposition 2.7. □

REMARK 2.10. It follows directly from the definition (2.25) that if the symbol $\sigma \in \Sigma_0^\mu$ [resp. $a \in \Gamma_0^\mu$] is supported in $\mathbb{R}^d \times \{\Lambda(\zeta) \le R\}$, then, for all u, the spectrum of $\mathrm{Op}^\gamma(\sigma)u$ [resp. $T_a^{\psi,\gamma} u$] is contained in $\{\Lambda(\zeta) \le 2R\}$. Similarly, if a is supported in $\{\Lambda(\zeta) \ge R\}$, then the spectrum of $T_a^{\psi,\gamma} u$ is contained in $\{\Lambda(\zeta) \le (1-\delta_1)^{-1} R\}$ with δ_1 as in (2.18).

2.1.3. *Paraproducts.* A function $a(x) \in L^\infty$ can be seen as a symbol in Γ_0^0, independent of (ξ,γ). With ψ given by (2.20) the symbol (2.22) associated to a is

$$\sigma_a(x,\xi,\gamma) = \sum_{k\ge 0} (\dot S_{k-N} a)(x) \left(\chi_k(\xi,\gamma) - \chi_{k-1}(\xi,\gamma)\right)$$

and the associated operator is

$$T_a^\gamma\, u := T_a^{\psi,\gamma} u \;=\; \sum_{k\ge 0} \dot S_{k-N} a\, \Delta_k^\gamma\, u\,. \tag{2.27}$$

THEOREM 2.11. *i) For all $a \in L^\infty$, T_a^γ is of order ≤ 0.*

ii) There is a constant C such that for all $a \in W^{1,\infty}$ and all u in the Schwartz class $\mathcal{S}(\mathbb{R}^d)$:

$$\|au - T_a^\gamma u\|_{1,\gamma} \;\le\; C\, \|\nabla a\|_{L^\infty} \|u\|_{0,\gamma}, \tag{2.28}$$

$$\gamma \|au - T_a^\gamma u\|_{0,\gamma} \;\le\; C\, \|\nabla a\|_{L^\infty} \|u\|_{\frac{p}{p_0}-1,\gamma}, \tag{2.29}$$

$$\|a\partial_j u - T_a^\gamma \partial_j u\|_{0,\gamma} \;\le\; C\, \|\nabla a\|_{L^\infty} \|u\|_{\frac{p}{p_j}-1,\gamma}. \tag{2.30}$$

PROOF. The first statement is clear from Proposition 2.9.

a) Because $[\partial_j, T_a^\gamma] = T^\gamma_{\partial_j a}$, $p/p_j - 1 \ge 0$, and

$$\|(\partial_j a) u\|_{L^2} + \|T^\gamma_{\partial_j a} u\|_{L^2} \lesssim \|(\partial_j a)\|_{L^\infty} \|u\|_{L^2}\,, \tag{2.31}$$

the estimate (2.30) follows from

$$\|\partial_j(au - T_a^\gamma u)\|_{L^2} \;\le\; C\, \|\nabla a\|_{L^\infty} \|u\|_{\frac{p}{p_j}-1,\gamma}\,. \tag{2.32}$$

Thus we prove (2.28), (2.29) and (2.32). Start from the identity

$$R(u) := au - T_a^\gamma u \;=\; \sum_{k>-N} \dot\Delta_k a\, S^\gamma_{k+N-1} u\,.$$

We first consider

$$R_1(u) = \sum_{k>-N} v_k\,, \quad v_k := \dot\Delta_k a \Big(\sum_{|l-k|<N} \Delta_l^\gamma u \Big).$$

Propositions 2.1 and 2.3 imply

$$\|v_k\|_{L^2} \lesssim 2^{-k}\, \|\nabla a\|_{L^\infty}\, 2^{k(1-q)} \varepsilon_k\, \|u\|_{q-1,\gamma}$$

where $\sum \varepsilon_k^2 \le 1$. Moreover, the spectrum of $\dot\Delta_k a$ is contained in the ball $\{\langle \xi' \rangle \le 2^{k+1}\}$, while the spectrum of $\Delta_l^\gamma u$ is contained in the set $\{\Lambda(\xi'',\gamma) \le 2^{l+1}\}$. Therefore the spectrum of v_k is contained in $\{\Lambda(\xi,\gamma) \le 2^{k+N+2}\}$. Thus, Proposition 2.2 implies that for $q > 0$

$$\|R_1 u\|_{q,\gamma} \lesssim \|\nabla a\|_{L^\infty}\, \|u\|_{q-1,\gamma}\,, \qquad q > 0. \tag{2.33}$$

Because

$$\gamma \le \Lambda(\xi,\gamma)^{p/p_0} \quad \text{and} \quad |\xi_j| \le \Lambda(\xi,\gamma)^{p/p_j},$$

we conclude that

$$(2.34)\qquad \begin{aligned} \|R_1(u)\|_{L^2} \le \|R_1(u)\|_{1,\gamma} &\lesssim \|\nabla a\|_{L^\infty} \|u\|_{0,\gamma}, \\ \gamma\|R_1(u)\|_{L^2} \le \|R_1(u)\|_{\frac{p}{p_0},\gamma} &\lesssim \|\nabla a\|_{L^\infty} \|u\|_{\frac{p}{p_0}-1,\gamma}, \\ \|\partial_j R_1(u)\|_{L^2} \le \|R_1(u)\|_{\frac{p}{p_j},\gamma} &\lesssim \|\nabla a\|_{L^\infty} \|u\|_{\frac{p}{p_j}-1,\gamma}. \end{aligned}$$

b) It remains to prove similar estimates for

$$R_2(u) = \sum_{k>-N} \dot\Delta_k a\, S^\gamma_{k-N} u\,.$$

Since $N \ge 3$, the spectrum of $w'_k := \dot\Delta_k a\, S^\gamma_{k-N} u$ is contained in the set $\{2^{k-2} \le \Lambda(\xi,\gamma) \le 2^{k+2}\}$. Moreover,

$$\|\Delta^\gamma_l u\|_{L^2} \lesssim 2^{l(1-q)} \varepsilon_l \|u\|_{q-1,\gamma}$$

with $\sum \varepsilon_l^2 \le 1$. Therefore,

$$\|w'_k\|_{L^2} \lesssim \|\nabla a\|_{L^\infty} \|u\|_{q-1,\gamma} 2^{-kq} \widetilde\varepsilon_k \quad \text{with } \widetilde\varepsilon_k = \sum_{l \le k-N} 2^{(l-k)(1-q)} \varepsilon_l\,.$$

If $q < 1$, $\sum \widetilde\varepsilon_k^2 \lesssim \sum \varepsilon_l^2$ and Proposition 2.2 implies that

$$(2.35)\qquad \|R_2(u)\|_{q,\gamma} \lesssim \|\nabla a\|_{L^\infty} \|u\|_{q-1,\gamma}, \qquad q < 1\,.$$

c) For $j > 0$, we write $\partial_j R_2(u) = R_2(\partial_j u) + R'_j$

$$R'_j = \sum_{k>-N} \dot\Delta_k \partial_j a\, S^\gamma_{k-N} u\,.$$

Suppose that we have proved the estimate

$$(2.36)\qquad \|R'_j\|_{L^2} \lesssim \|\nabla a\|_{L^\infty} \|u\|_{0,\gamma}\,.$$

Applying (2.35) with $q = 0$, we obtain that

$$\|\gamma R_2(u)\|_{L^2} \lesssim \|\nabla a\|_{L^\infty} \|\gamma u\|_{-1,\gamma} \lesssim \|\nabla a\|_{L^\infty} \|u\|_{\frac{p}{p_0}-1,\gamma}$$

With (2.34), this implies (2.29). Similarly, (2.35) implies that

$$\|R_2(\partial_j u)\|_{L^2} \lesssim \|\nabla a\|_{L^\infty} \|\partial_j u\|_{-1,\gamma} \lesssim \|\nabla a\|_{L^\infty} \|u\|_{\frac{p}{p_j}-1,\gamma}$$

With (2.36), this implies that

$$\|\partial_j R_2(u)\|_{L^2} \lesssim \|\nabla a\|_{L^\infty} \|u\|_{\frac{p}{p_j}-1,\gamma}$$

and with (2.34), the estimates (2.32) and (2.30) follow. This result still holds for $j = 0$, replacing $\partial_j u$ by γu in the estimates above. With the second estimate in (2.34) this implies (2.29).

Next, we note that

$$1 + \gamma\Lambda^{1-p/p_0} + \sum |\zeta_j| \Lambda^{1-p/p_j} \approx \Lambda$$

implying

$$(2.37)\qquad \|R_2(u)\|_{1,\gamma} \lesssim \|R_2(u)\|_{0,\gamma} + \gamma\|R_2(u)\|_{1-\frac{p}{p_0},\gamma} + \sum \|\partial_j R_2(u)\|_{1-\frac{p}{p_j},\gamma}\,.$$

The first term is controlled by (2.35) which implies

$$\|R_2(u)\|_{0,\gamma} \lesssim \|\nabla a\|_{L^\infty} \|u\|_{-1,\gamma} \lesssim \|\nabla a\|_{L^\infty} \|u\|_{0,\gamma}\,.$$

Similarly, (2.35) implies that

$$\|\gamma R_2(u)\|_{1-\frac{p}{p_0},\gamma} \lesssim \|\nabla a\|_{L^\infty}\|\gamma u\|_{-\frac{p}{p_0},\gamma} \lesssim \|\nabla a\|_{L^\infty}\|u\|_{0,\gamma}$$

$$\|R_2(\partial_j u)\|_{1-\frac{p}{p_j},\gamma} \lesssim \|\nabla a\|_{L^\infty}\|\partial_j u\|_{-\frac{p}{p_j},\gamma} \lesssim \|\nabla a\|_{L^\infty}\|u\|_{0,\gamma}$$

Together with (2.36), the later inequality implies

$$\|\partial_j R_2(u)\|_{1-\frac{p}{p_j},\gamma} \lesssim \|\nabla a\|_{L^\infty}\|u\|_{0,\gamma}$$

Therefore, we have proved that each term in the right hand side of (2.37) is dominated by the right hand side of (2.28), which finishes the proof of (2.28).

d) It only remains to prove (2.36). The spectrum of $w''_k := \dot\Delta_k \partial_j a\, S^\gamma_{k-N} u$ is contained in the set $\{2^{k-2} \leq \Lambda(\xi,\gamma) \leq 2^{k+2}\}$. Hence,

$$\|R'_j\|^2_{L^2} \lesssim \sum_{k>-N} \|w''_k\|^2_{L^2}\,.$$

Thus, the inequality stated in the next proposition implies (2.36), finishing the proof of Theorem 2.11. □

PROPOSITION 2.12. *There is a constant C such that for all $b \in L^\infty$, $u \in L^2$ and $\gamma \geq 0$:*

$$\sum_{k>-N} \|\dot\Delta_k b\; S^\gamma_{k-N} u\|^2_{L^2} \leq C\,\|b\|^2_{L^\infty}\,\|u\|^2_{L^2}\,. \tag{2.38}$$

This is a classical result from Harmonic Analysis (see [**Co-Me**], [**St**], at least in the homogeneous case), based on the fact that $\sum_k |\dot\Delta_k b(x)|^2 \otimes \delta_{t=2^{-k}}$ is a Carleson measure, which is true if $b \in BMO$. For the convenience of the reader, and to cover the quasi-homogeneous case, we sketch a proof of (2.38) in the easier case $b \in L^\infty$. The first step is the following.

LEMMA 2.13. *There is a constant C such that for all $b \in L^\infty(\mathbb{R}^d)$ and all open set $\Omega \subset \mathbb{R}^d$:*

$$\sum_{k>-N} \|\dot\Delta_k b\|^2_{L^2(\Omega_k)} \leq C\ \mathrm{meas}(\Omega)\,\|b\|^2_{L^\infty}\,, \tag{2.39}$$

where Ω_k denotes the set of points $x \in \Omega$ such that the ball $B(x, 2^{-k}) := \{y \in \mathbb{R}^d : \langle x-y\rangle < 2^{-k}\}$ is contained in Ω.

PROOF. Write $b = b' + b''$ with $b' = b1_\Omega$. Denote by $I(b)$ the left hand side of (2.39). Then $I(b) \leq 2I(b') + 2I(b'')$. Therefore, it is sufficient to prove the inequality separately for b' and b''. One has

$$\sum_{k>-N} \|\dot\Delta_k b'\|^2_{L^2(\Omega_k)} \leq \sum_{k>-N} \|\dot\Delta_k b'\|^2_{L^2(\mathbb{R}^d)} \leq \|b'\|^2_{L^2} \leq \|b\|^2_{L^\infty}\,\mathrm{meas}(\Omega)\,.$$

Thus, it remains to prove (2.39) for b''.

The kernel of $\dot\Delta_k$ is $G_k(x) = 2^{kD} G_0(2^k \cdot x)$ where G_0 belongs to the Schwartz class $\mathcal{S}$. Thus

$$\dot\Delta_k b''(x) = \int 2^{kD} G_0(2^k \cdot (x-y)) b''(y)\,dy\,.$$

On the support of b'', $y \notin \Omega$ and for $x \in \Omega_l$, the distance $\langle x-y\rangle$ is larger than 2^{-l}. Thus, for $x \in \Omega_l$

$$|\dot\Delta_k b''(x)| \leq \|b''\|_{L^\infty} \int_{\{\langle y\rangle \geq 2^{-l}\}} 2^{kD}\,|G_0(2^k \cdot y)|\,dy = \|b''\|_{L^\infty}\, g^*_{k-l}$$

with

$$g_l^* = \int_{\{\langle y\rangle \geq 2^l\}} |G_0(y)|\, dy.$$

Let $\Omega'_{-N} := \Omega_{-N}$ and for $l > -N$, let $\Omega'_l = \Omega_l \setminus \Omega_{l-1}$. Then the pointwise estimate above implies that

$$\|\dot\Delta_k b''\|^2_{L^2(\Omega'_l)} \leq \|b\|^2_{L^\infty} \operatorname{meas}(\Omega'_l)\, \left(g^*_{k-l}\right)^2 . \tag{2.40}$$

Since $\Omega_k = \bigcup_{l\leq k} \Omega'_l$,

$$\sum_{k>-N} \|\dot\Delta_k b''\|^2_{L^2(\Omega_k)} = \sum_{k>-N} \sum_{l=-N}^{k} \|\dot\Delta_k b''\|^2_{L^2(\Omega'_l)} .$$

Using (2.40), this implies

$$\sum_{k>-N} \|\dot\Delta_k b''\|^2_{L^2(\Omega_k)} \leq \sum_{l\geq -N} \sum_{k\geq l} \|b\|^2_{L^\infty} \left(g^*_{k-l}\right)^2 \operatorname{meas}(\Omega'_l) .$$

Since $G_0 \in \mathcal{S}$, the sequence g^*_k is rapidly decreasing and thus in $\ell^2(\mathbb{N})$. Therefore,

$$\sum_{k>-N} \|\dot\Delta_k b''\|^2_{L^2(\Omega_k)} \lesssim \|b\|^2_{L^\infty} \sum_{l\geq -N} \operatorname{meas}(\Omega'_l) = \|b\|^2_{L^\infty} \operatorname{meas}(\Omega) .$$

□

COROLLARY 2.14. *There is a constant C such that for all $b \in L^\infty(\mathbb{R}^d)$ and all sequence v_k in L^2, one has*

$$\sum_{k>-N} \|(\dot\Delta_k b)\, v_k\|^2_{L^2} \leq C\, \|b\|^2_{L^\infty}\, \|v_*\|^2_{L^2} , \tag{2.41}$$

where

$$v_*(x) = \sup_{k>-N} \sup_{B(x,2^{-k})} |v_k(y)| . \tag{2.42}$$

PROOF. Let $b_k = \dot\Delta_k b$. Then

$$\|b_k v_k\|^2_{L^2} = 2\int_0^\infty \lambda \|b_k\|^2_{L^2(U_k(\lambda))}\, d\lambda , \quad \text{where} \quad U_k(\lambda) = \{|v_k| > \lambda\}.$$

For $\lambda > 0$, let $\Omega(\lambda) = \{|v_*| > \lambda\}$. This is the set of points x such that there are $k > -N$ and y such that $\langle x - y\rangle < 2^{-k}$ and $|v_k(y)| > \lambda$. Thus $\Omega(\lambda)$ is open and if $|v_k(y)| > \lambda$, the ball $B(y, 2^{-k})$ is contained in $\Omega(\lambda)$. This shows that for all k, $U_k(\lambda) \subset \Omega_k(\lambda)$, where the Ω_k's are defined as in Lemma 2.13. Thus

$$\sum_{k>-N} \|b_k\|^2_{L^2(U_k(\lambda))} \leq \sum_{k>-N} \|b_k\|^2_{L^2(\Omega_k(\lambda))} \lesssim \|b\|^2_{L^\infty} \operatorname{meas}(\Omega(\lambda)) ,$$

and

$$\sum_{k>-N} \|b_k v_k\|^2_{L^2} \lesssim \|b\|^2_{L^\infty} \int_0^\infty 2\lambda \operatorname{meas}(\Omega(\lambda))\, d\lambda = \|b\|^2_{L^\infty}\, \|v_*\|^2_{L^2} .$$

□

LEMMA 2.15. *There is a constant C such that for all $u \in L^2$ and all $\gamma \geq 0$, the function v_* defined by (2.42) and $v_k = S^\gamma_{k-N} u$ satisfies*

$$v_*(x) \leq C u^*(x) := \sup_{r>0} \frac{1}{\operatorname{meas}(B(x,r))} \int_{B(x,r)} |u(y)| dy . \tag{2.43}$$

PROOF. S_k^γ is a convolution operator with kernel $G_k(\cdot,\gamma)$ equal to the inverse Fourier transform of $\xi \mapsto \chi(2^{-k}\Lambda(\xi,\gamma))$ Note that

$$2^{-k}\Lambda(\xi,\gamma) = \Big(2^{-2kp} + \langle 2^{-k}\cdot(\xi,\gamma)\rangle^{2p}\Big)^{\frac{1}{2p}}.$$

Thus,

$$G_k(x,\gamma) = 2^{kD} H_k(2^k\cdot x, 2^{-kp/p_0}\gamma).$$

where $H_k(\cdot,\gamma')$ is the inverse Fourier transform of $\chi\big((2^{-kp} + \langle\cdot,\gamma'\rangle^{2p})^{1/2p}\big)$. It vanishes when γ' is large, more precisely when $\langle 0,\gamma'\rangle > 2$. For $\gamma' \le 2$, the $H_k(\cdot,\gamma')$ remain in a bounded set of the Schwartz' space $\mathcal{S}$. Thus

$$G_l^* := \sup_{k,\gamma'}\sup_{A_l} |H_k(x,\gamma')|,$$

where A_l is the annulus $\{2^{l-1} \le \langle x\rangle \le 2^l\}$ for $l > 0$ and the ball $\{\langle x\rangle \le 1\}$ for $l = 0$, is rapidly decreasing. In particular

$$\sum_{l\ge 0} 2^{lD} G_l^* < \infty. \tag{2.44}$$

With $\gamma' = 2^{(N-k)p/p_0}\gamma$, one has

$$|v_k(x-x')| \le 2^{(k-N)D}\int |H_k(2^{k-N}\cdot y,\gamma')|\,|u(x-x'-y)|dy.$$

Splitting the domain of integration into the union over l of the $A_l^k = \{y : 2^{k-N}\cdot y \in A_l\}$, yields

$$|v_k(x-x')| \le 2^{(k-N)D}\sum_{l\ge 0} G_l^*\int_{A_l^k} |u(x-x'-y)|dy.$$

When $y \in A_l^k$, $\langle y\rangle \le 2^{l-k+N}$. If in addition $\langle x-x'\rangle \le 2^{-k} \le 2^{l-k+N}$, then $x'+y$ belongs to the ball centered at x of radius $2^{-k+l+N+1}$ whose measure is equal to $c_0 2^{(l-k+N+1)D}$ (see (2.6)). Thus

$$\int_{A_l^k} |u(x-x'-y)|dy \le c_0 2^{(l-k+N+1)D}\, u^*(x)$$

and therefore

$$\sup_{|x'|\le 2^{-k}} |v_k(x-x')| \;\le\; 2c_0\sum_{l\ge 0} 2^{lD} G_l^*\; u^*(x).$$

With (2.44), the estimate (2.43) follows. □

PROOF OF PROPOSITION 2.12. The Corollary 2.14 and Lemma 2.15 imply that

$$\sum_{k>-N} \|\dot\Delta_k b S_{k-N}^\gamma u\|_{L^2}^2 \le C\,\|b\|_{L^\infty}^2\,\|u^*\|_{L^2}^2.$$

That $\|u^*\|_{L^2} \le C\|u\|_{L^2}$ is a general fact about maximal functions (see e.g. [**St**], [**Co-Me**]) which only uses the properties (2.6) (2.7). □

2.1.4. *Symbolic calculus.*

THEOREM 2.16. *Consider $a \in \Gamma_1^{\mu}$ and $b \in \Gamma_1^{m'}$. Then $ab \in \Gamma_1^{\mu+\mu'}$ and $T_a^\gamma \circ T_b^\gamma - T_{ab}^\gamma$ is of order $\leq \mu + \mu' - 1$. More precisely, one has*

$$\|(T_a^\gamma T_b^\gamma - T_{ab}^\gamma)u\|_{s,\gamma} \leq C \Big(\|a\|_{(\mu,N)} \|\nabla_x b\|_{(\mu',N)} + \|\nabla_x a\|_{(\mu,N)} \|b\|_{(\mu',N)} \Big) \|u\|_{s+\mu+\mu'-1,\gamma}$$

where C and N depend only on the indices s and μ.

This extends to matrix valued symbols and operators.

Moreover, if b is independent of x, then $T_a^\gamma \circ T_b^\gamma = T_{ab}^\gamma$.

REMARK 2.17. *The definition of the operators T_a^γ involves the choice of admissible functions ψ. Proposition 2.9 implies that the result does not depend on the particular choice of ψ. To be precise, in the statements below, we consider that the quantification T is associated to a fixed given admissible function ψ, for instance (2.20) with $N = 3$. However, within the proofs, we use other functions ψ, at the price of error terms controlled by Proposition 2.9.*

PROOF. **a)** The last statement of the theorem is clear using the definitions, since T_b^γ is just the action of the Fourier multiplier $b(\zeta)$. We now focus on the main part of the theorem.

b) We first consider two symbols $\sigma_1 \in \Sigma_1^\mu$ and $\sigma_2 \in \Sigma_1^{\mu'}$ satisfying the spectral condition in Definition 2.4 with a parameter $\delta < 1/3$. Using (2.24) and (2.25), one gets that $\mathrm{Op}^\gamma(\sigma_1) \circ \mathrm{Op}^\gamma(\sigma_2) = \mathrm{Op}^\gamma(\sigma)$ with

$$\begin{aligned} \sigma(x,\xi,\gamma) &= \frac{1}{(2\pi)^d} \int e^{ix(\eta-\xi)}\, \sigma_1(x,\eta,\gamma)\, \widehat{\sigma}_2^1(\eta-\xi,\xi,\gamma)\, d\eta \\ &= \frac{1}{(2\pi)^d} \int e^{ix\eta}\, \sigma_1(x,\xi+\eta,\gamma)\, \widehat{\sigma}_2^1(\eta,\xi,\gamma)\, d\eta\,. \end{aligned}$$

Thus

$$\widehat{\sigma}^1(\eta,\xi,\gamma) = \frac{1}{(2\pi)^d} \int \widehat{\sigma}_1^1(\eta-\eta',\xi+\eta',\gamma)\, \widehat{\sigma}_2^1(\eta',\xi,\gamma)\, d\eta'\,.$$

On the support of the integral, $\langle \eta - \eta' \rangle \leq \delta\Lambda(\xi+\eta',\gamma)$ and $\langle \eta' \rangle \leq \delta\Lambda(\xi,\gamma)$. Thus $\langle \eta \rangle \leq (1+\delta)\langle \eta' \rangle + \delta\Lambda(\xi,\gamma) \leq (2\delta+\delta^2)\Lambda(\xi,\gamma)$ and σ satisfies the spectral property in *iii)* of Definition 2.4, since $\delta < 1/3$.

Since σ_2 satisfies the spectral property, there is an admissible function θ such that $\theta\widehat{\sigma}_2^1 = \widehat{\sigma}_2^1$. Inserting θ in the definition of σ yields

$$\sigma(x,\xi,\gamma) = \int H(x,y,\xi,\gamma)\, \sigma_2(y,\xi,\gamma)\, dy$$

with

$$H(x,y,\xi,\gamma) := \frac{1}{(2\pi)^d} \int e^{i(x-y)\eta} \sigma_1(x,\xi+\eta,\gamma) \theta(\eta,\xi,\gamma)\, d\eta\,.$$

Next, we use Taylor's formula,

$$\sigma_1(x,\xi+\eta,\gamma) = \sigma_1(x,\xi,\gamma) + \sum \sigma_{1,j}(x,\eta,\xi,\gamma)\eta_j$$

with

$$\sigma_{1,j}(x,\eta,\xi,\gamma) = \int_0^1 \partial_{\xi_j} \sigma_1(x,\xi+t\eta,\gamma) dt\,.$$

Note that here, $t\eta$ is the usual multiplication of η by t, not the dilation $t \cdot \eta$.

The first term contributes in σ to $\sigma_1(x,\xi,\gamma)\sigma_2(x,\xi,\gamma)$. The remainder $r := \sigma - \sigma_1\sigma_2$ satisfies

$$r(x,\xi,\gamma) = \sum_{j=1}^{d} \int G_j(x, x-y, \xi, \gamma)\, (\partial_{x_j}\sigma_2)(y,\xi,\gamma)\, dy\,,$$

$$G_j(x,y,\xi,\gamma) = \frac{-i}{(2\pi)^d} \int e^{iy\eta} \sigma_{1,j}(x,\eta,\xi,\gamma)\theta(\eta,\xi,\gamma)\, d\eta\,.$$

For (η,ξ,γ) in the support of θ and for $t \in [0,1]$, $\langle t\eta\rangle \le \langle \eta \rangle \le \delta\Lambda(\xi,\gamma)$ and thus $\Lambda(\xi + t\eta, \gamma) \approx \Lambda(\xi,\gamma)$. With the estimates (2.19) for θ and (2.16) for σ_1, this implies that

$$|\partial_\xi^\alpha \partial_\eta^\beta (\sigma_{1,j}\theta)(\eta,\xi,\gamma)| \le C_{\alpha,\beta}\Lambda(\xi,\gamma)^{\mu - \langle\alpha\rangle - \langle\beta\rangle - p/p_j}\,.$$

(Recall that $\sigma_{1,j}$ involves the derivative $\partial_{\xi_j}\sigma_1$, yielding the extra factor p/p_j in the estimate above). As for (2.21), these estimates imply

$$\|\partial_\xi^\alpha G_j(x,\cdot,\xi,\gamma)\|_{L^1(\mathbb{R}^d)} \le C_\alpha \Lambda^{\mu-1-\langle\alpha\rangle}\,.$$

We have used that $p/p_j \ge 1$. Together with the estimates (2.16) for $\partial_{x_j}\sigma_2$, this shows that $r \in \Gamma_0^{\mu+\mu'-1}$. More precisely, there is N', such that for all N, there is C such that

$$\|r\|_{(\mu+\mu'-1,N)} \le C\, \|\sigma_1\|_{(\mu,N+N')}\, \|\nabla_x\sigma_2\|_{(\mu',N)}. \tag{2.45}$$

Because σ and $\sigma_1\sigma_2$ both satisfy the spectral condition, we conclude that $r \in \Sigma_0^{\mu+\mu'-1}$.

c) Consider $a \in \Gamma_1^\mu$ and $b \in \Gamma_1^{\mu'}$. By Proposition 2.9, changing the admissible function in the definition of T^γ changes T_a^γ and T_b^γ by operators of order $\mu - 1$ and $\mu' - 1$, with norms controlled by the semi-norms of $\nabla_x a$ and $\nabla_x b$. This does not alter the result. Therefore, changing ψ (for instance increasing N in (2.20)) if necessary, we can assume that the parameter δ_2 occurring in (2.18) is small enough. In this case, the associated symbols $\sigma_a \in \Sigma_1^\mu$ and $\sigma_b \in \Sigma_1^{\mu'}$ satisfy the spectral condition with a parameter $\delta < 1/3$. Therefore, $T_a^\gamma \circ T_b^\gamma = \mathrm{Op}^\gamma(\sigma_a) \circ \mathrm{Op}^\gamma(\sigma_b) = \mathrm{Op}^\gamma(\sigma_a\sigma_b) + \mathrm{Op}^\gamma(r)$ with $r \in \Sigma_0^{\mu+\mu'-1}$.

On the other hand, Proposition 2.7 implies that $a - \sigma_a \in \Gamma_0^{\mu-1}$, $b - \sigma_b \in \Gamma_0^{\mu'-1}$ and $ab - \sigma_{ab} \in \Gamma_0^{\mu+\mu'-1}$. Thus, $r' = \sigma_a\sigma_b - \sigma_{ab} \in \Sigma_0^{\mu+\mu'-1}$. Moreover, the norms of r' involve bounds of $\nabla_x a$ or $\nabla_x b$. One has $T_{ab}^\gamma = \mathrm{Op}^\gamma(\sigma_{ab}) = \mathrm{Op}^\gamma(\sigma_a\sigma_b) - \mathrm{Op}^\gamma(r')$, and thus $T_a^\gamma \circ T_b^\gamma - T_{ab}^\gamma = \mathrm{Op}^\gamma(r + r')$, which is of order $\le \mu + \mu' - 1$ and the theorem follows. ☐

Similarly, the next two theorems are extensions of known results ([**Bo**], [**Mey**]) to the framework of quasi-homogeneous symbols.

THEOREM 2.18. *Consider a matrix valued symbol* $a \in \Gamma_1^\mu$. *Denote by* $(T_a^\gamma)^*$ *the adjoint operator of* T_a^γ *and by* $a^*(x,\xi,\gamma)$ *the adjoint of the matrix* $a(x,\xi,\gamma)$. *Then* $(T_a^\gamma)^* - T_{a^*}^\gamma$ *is of order* $\le \mu - 1$. *More precisely, one has*

$$\|((T_a^\gamma)^* - T_{a^*}^\gamma)u\|_{s,\gamma} \le C\|\nabla_x a\|_{(\mu,N)}\, \|u\|_{s+\mu-1,\gamma}$$

where C *and* N *only depend on the indices* s *and* μ.

PROOF. It is sufficient to consider scalar symbols.

a) Consider $\sigma \in \Sigma_1^\mu$ with parameter $\delta < 1/2$. On the Fourier side, the kernel of $\mathrm{Op}^\gamma(\sigma)$ is $\widehat{\sigma}^1(\xi - \eta, \eta, \gamma)$ (see (2.25)). Thus the kernel of its adjoint is $\overline{\widehat{\sigma}^1(\eta - \xi, \xi, \gamma)} = \widehat{\overline{\sigma}}^1(\xi - \eta, \xi, \gamma)$. Therefore, the adjoint $(\mathrm{Op}^\gamma(\sigma))^*$ is the operator $\mathrm{Op}^\gamma(\widetilde{\sigma})$ with symbol $\widetilde{\sigma}$ defined by $\widehat{\widetilde{\sigma}}^1(\xi - \eta, \eta, \gamma) = \widehat{\overline{\sigma}}^1(\xi - \eta, \xi, \gamma)$, that is

$$\widehat{\widetilde{\sigma}}^1(\eta, \xi, \gamma) = \widehat{\overline{\sigma}}^1(\eta, \xi + \eta, \gamma)\,.$$

On the support of $\widehat{\widetilde{\sigma}}^1$, $\langle \eta \rangle \le \delta \Lambda(\xi + \eta, \gamma) \le \delta(\Lambda(\xi, \gamma) + \langle \eta \rangle)$, and therefore $\langle \eta \rangle \le \frac{\delta}{1-\delta}\Lambda(\xi, \gamma)$. Since $\delta < 1/2$, the spectral condition is satisfied.

Since both $\widehat{\sigma}^1(\eta, \xi, \gamma)$ and $\widehat{\overline{\sigma}}^1(\eta, \xi + \eta, \gamma)$ satisfy the spectral condition, there is an admissible cut-off function θ such that $\theta\widehat{\sigma}^1 = \widehat{\sigma}^1$ and $\theta\widehat{\overline{\sigma}}^1 = \widehat{\overline{\sigma}}^1$. Then

$$\widehat{\overline{\sigma}}^1(\eta, \xi + \eta, \gamma) = \widehat{\overline{\sigma}}^1(\eta, \xi, \gamma) + \sum_{j=1}^{d} \theta(\eta, \xi, \gamma) \int_0^1 \eta_j\, \partial_{\xi_j} \widehat{\overline{\sigma}}^1(\eta, \xi + t\eta, \gamma) dt\,.$$

Taking the inverse Fourier transform with respect to the first set of variables, we obtain that $\widetilde{\sigma} = \overline{\sigma} + r$, where

$$r(x, \xi, \gamma) = \sum_j \frac{1}{(2\pi)^d} \int e^{i(x-y)\eta} \rho_j(y, \eta, \xi, \gamma)\, dy\, d\eta\,,$$

and

$$\rho_j(x, \eta, \xi, \gamma) = -i\theta(\eta, \xi, \gamma) \int_0^1 \partial_{x_j} \partial_{\xi_j} \overline{\sigma}(x, \xi + t\eta, \gamma)\, dt\,.$$

As in the proof of Theorem 2.16, $\langle \eta \rangle \delta \le \Lambda(\xi, \gamma)$ and $\Lambda(\xi + t\eta, \gamma) \approx \Lambda(\xi, \varphi)$ on the support of $\theta(\eta, \xi, \gamma)\sigma(x, \xi + t\eta, \gamma)$ and for $|\alpha| + |\beta \le N$,

$$|\partial_\xi^\alpha \partial_\eta^\beta \rho_j(x, \eta, \xi, \gamma)| \le C \|\nabla_x \sigma\|_{(\mu, N+1)}\, \Lambda(\xi, \gamma)^{\mu - \langle \alpha \rangle - \langle \beta \rangle - p/p_j}\,.$$

Therefore, the integrals

$$G_j(x, y, \xi, \gamma) = \frac{1}{(2\pi)^d} \int e^{ix\eta}\, \rho_j(y, \eta, \xi, \gamma)\, d\eta$$

satisfy for all α and N':

$$|\partial_\xi^\alpha G_j(x, y, \xi, \gamma)| \le C \|\nabla_x \sigma\|_{(\mu, N+1+N')} \frac{\Lambda^D}{(1 + \langle \Lambda \cdot x \rangle)^N} \Lambda^{\mu - \langle \alpha \rangle - 1}$$

with $\Lambda = \Lambda(\xi, \gamma)$. Thus, with N' large enough

$$\|\partial_\xi^\alpha G_j(x - \cdot, \cdot, \xi, \gamma)\|_{L^1(\mathbb{R}^d)} \le C_\alpha \|\nabla_x \sigma\|_{(\mu, N')}\, \Lambda(\xi, \gamma)^{\mu - \langle \alpha \rangle - 1}\,.$$

This implies that $r \in \Gamma_0^{\mu-1}$, and, because both $\overline{\sigma}$ and $\widetilde{\sigma}$ satisfy the spectral property, $r \in \Sigma_0^{\mu-1}$ and $\mathrm{Op}^\gamma(r) = (\mathrm{Op}^\gamma(\sigma))^* - \mathrm{Op}^\gamma(\widetilde{\sigma})$ is of order $\mu - 1$, with bounds depending on the semi-norms of $\nabla_x \sigma$ in Γ_0^μ.

b) Consider now $a \in \Gamma_1^\mu$. Changing the admissible function ψ in the definition of T^γ, changes T_a^γ by an operator R of order $\le \mu - 1$. The adjoint R^* is also bounded from H^s to $\mathrm{H}^{s-\mu+1}$ for all $s \in \mathbb{R}$. Moreover, the norms only depend on semi-norms of $\nabla_x a$. Thus, to prove Theorem (2.18), we can assume that the parameters δ of the admissible cut-off ψ are smaller than $1/2$. In this case, the analysis in a) applies to σ_a and the theorem follows. □

THEOREM 2.19. *Consider a $N \times N$ matrix symbol $a \in \Gamma_1^\mu$ and a $N \times M$ matrix symbol $w \in \Gamma_1^0$. Suppose that there is a scalar real symbol $\chi \in \Gamma_1^0$ and $c > 0$ such that $\chi \geq 0$, $\chi w = w$ and*

$$\forall (x, \xi, \gamma) : \quad \chi^2(x, \xi, \gamma) \operatorname{Re} a(x, \xi, \gamma) \geq c\chi^2(x, \xi, \gamma)\, \Lambda^\mu(\xi, \gamma).$$

Then, there are C and N such that for all $u \in \mathrm{H}^{\mu/2}$ and all $\gamma \geq 0$,

$$\frac{c}{2} \|T_w^\gamma u\|^2_{\frac{\mu}{2},\gamma} \leq \operatorname{Re} \big(\!\big(T_a^\gamma T_w^\gamma u, T_w^\gamma u \big)\!\big)_{L^2} \; + \; C\, K^2 \|u\|^2_{\frac{\mu}{2}-1,\gamma} \tag{2.46}$$

where

$$K = \|\nabla_x a\|_{(\mu,N)} + \|\nabla_x \chi\|_{(0,N)} + \|\nabla_x w\|_{(0,N)}$$

and C is bounded when a, χ and w remain in bounded sets.

Here, $(\!(\cdot, \cdot)\!)$ denotes the scalar product in $L^2(\mathbb{R}^d)$, which can be extended as the duality pairing $\mathrm{H}^s \times \mathrm{H}^{-s}$.

PROOF. The assumption implies that $\operatorname{Re} a - \frac{3c}{4}\Lambda^\mu$ is positive definite on the support of χ. Therefore, one can define

$$b = b^* = \chi\big(\operatorname{Re} a - \frac{3c}{4}\Lambda^\mu\big)^{1/2} \in \Gamma_1^{\mu/2}.$$

Then,

$$\operatorname{Re} a = b^* b + \frac{3c}{4}\Lambda^\mu + a', \quad a' := (1 - \chi^2)(\operatorname{Re} a - \frac{3c}{4}\Lambda^\mu).$$

One has

$$\operatorname{Re} \big(\!\big(T_a^\gamma v, v \big)\!\big) = \frac{1}{2}\big(\!\big((T_a^\gamma + (T_a^\gamma)^*) v, v \big)\!\big).$$

The symbolic calculus implies that

$$\frac{1}{2}(T_a^\gamma + (T_a^\gamma)^*) = T^\gamma_{\operatorname{Re} a} + R_1^\gamma = (T_b^\gamma)^* T_b^\gamma + \frac{3c}{4} T^\gamma_{\Lambda^\mu} + T^\gamma_{a'} + R_2^\gamma$$

where R_1 and R_2 are families of order $\leq \mu - 1$. Hence,

$$\operatorname{Re} \big(\!\big(T_a^\gamma v, v \big)\!\big) = \|T_b^\gamma v\|^2_{L^2} + \frac{3c}{4} \|v\|^2_{\frac{\mu}{2},\gamma} + \operatorname{Re} \big(\!\big(T_{a'}^\gamma v, v \big)\!\big) + \operatorname{Re} \big(\!\big(R_2^\gamma v, v \big)\!\big).$$

We apply these identity to $v = T_w^\gamma u$. Note that $a'w = 0$ since $(1 - \chi^2)w = 0$. Thus, $T_{a'}^\gamma T_w^\gamma = R_3^\gamma$ is of order $\mu - 1$. This implies that

$$\frac{3c}{4} \|T_w^\gamma u\|^2_{\frac{\mu}{2},\gamma} \leq \operatorname{Re} \big(\!\big(T_a^\gamma T_w^\gamma u, T_w^\gamma u \big)\!\big) + M \|u\|_{\frac{\mu}{2}-1,\gamma} \|T_w^\gamma u\|_{\frac{\mu}{2},\gamma},$$

with $M = \|R_3^\gamma\| + \|R_2^\gamma T_w^\gamma\|$, where the norms are taken in the space of bounded operators from $\mathrm{H}^{\frac{\mu}{2}-1}$ to $\mathrm{H}^{-\frac{\mu}{2}}$. The symbolic calculus implies that

$$M \leq C\Big(\|\nabla_x a\|_{(\mu,N)} + \|\nabla_x b\|_{(\mu/2,N)} + \|\nabla_x \chi\|_{(0,N)} + \|\nabla_x w\|_{(0,N)} \Big)$$

for some N. Therefore $M \leq CK$ and the theorem follows. □

2.2. The semi-classical calculus. The semi-classical quantization associates to a symbol $\sigma(\widetilde{y}, \widetilde{\eta}, \gamma)$ the operator

$$\mathrm{Op}^{\varepsilon,\gamma}(\sigma)u(\widetilde{y}) := \frac{1}{(2\pi)^d} \int e^{i\widetilde{\eta}\cdot\widetilde{y}}\, \sigma(\widetilde{y}, \varepsilon\widetilde{\eta}, \varepsilon\gamma)\, \widehat{u}(\widetilde{\eta})\, d\widetilde{\eta}\,,$$

so that, if σ is independent of $\widetilde{y}$, the operator is defined by the Fourier multiplier $\sigma(\varepsilon\widetilde{\eta}, \varepsilon\gamma)$. Note that here $\varepsilon\widetilde{\eta}$ is the usual multiplication by ε, not the quasi-homogeneous dilation $\varepsilon \cdot \widetilde{\eta}$. An alternate definition is

$$\mathrm{Op}^{\varepsilon,\gamma}(\sigma) = (H_\varepsilon)^{-1}\mathrm{Op}^{\gamma'}(\sigma^\varepsilon)H_\varepsilon \quad \text{for } \gamma' = \varepsilon\gamma\,,$$

where $H_\varepsilon u = \varepsilon^{d/2}\, u(\varepsilon x)$ and $\sigma^\varepsilon(x, \xi, \gamma) = \sigma(\varepsilon x, \xi, \gamma)$. We extend this definition to the para-differential context.

DEFINITION 2.20. For a symbol $a \in \Gamma_0^\mu$, $\varepsilon > 0$ and $\gamma \geq 0$, $P_a^{\varepsilon,\gamma}$ is the operator defined by

$$P_a^{\varepsilon,\gamma} = (H_\varepsilon)^{-1} T_{a^\varepsilon}^{\gamma'} H_\varepsilon \quad \text{for } \gamma' = \varepsilon\gamma \text{ and } a^\varepsilon(x, \xi, \gamma) = a(\varepsilon x, \xi, \gamma)\,. \tag{2.47}$$

On H^s, we introduce the norms

$$\|u\|_{s,\varepsilon,\gamma} = \Big(\int \Lambda(\varepsilon\xi, \varepsilon\gamma)^{2s} |\widehat{u}(\xi)|^2\, d\xi \Big)^{1/2}. \tag{2.48}$$

We note that

$$\|u\|_{s,\varepsilon,\gamma} = \|H_\varepsilon u\|_{s,\gamma'}\,, \quad \text{with } \gamma' = \varepsilon\gamma\,. \tag{2.49}$$

When $a \in \Gamma_0^\mu$, the family $\{a^\varepsilon : \varepsilon \in]0,1]\}$ is bounded in Γ_0^μ. Therefore, by Proposition 2.9 there is a constant C such that for all $\varepsilon \in]0,1]$, all $\gamma' \geq 0$ and all $v \in \mathrm{H}^s$,

$$\|T_{a^\varepsilon}^{\gamma'} v\|_{s-\mu,\gamma'} \leq C\|v\|_{s,\gamma'}\,.$$

Applied to $v = H_\varepsilon u$ and $\gamma' = \varepsilon\gamma$, we deduce that

PROPOSITION 2.21. *For $a \in \Gamma_0^\mu$, there is C such that for all $\varepsilon \in]0,1]$, all $\gamma \geq 0$ and all $u \in \mathrm{H}^s$,*

$$\|P_a^{\varepsilon,\gamma} u\|_{s-\mu,\varepsilon,\gamma} \leq C\|u\|_{s,\varepsilon,\gamma}\,.$$

Next, we remark that when $a \in \Gamma_1^\mu$, the $\nabla_x a^\varepsilon$ are $O(\varepsilon)$ in Γ_0^μ. Thus, Theorem 2.16 implies that for $a \in \Gamma_1^\mu$ and $b \in \Gamma_1^{\mu'}$, one has

$$\|(T_{a^\varepsilon}^{\gamma'} T_{b^\varepsilon}^{\gamma'} - T_{a^\varepsilon b^\varepsilon}^{\gamma'})v\|_{s,\gamma'} \leq C\,\varepsilon\, \|v\|_{s+\mu+\mu'-1,\gamma'}.$$

Using (2.47) and (2.49), this implies the next result.

PROPOSITION 2.22. *For $a \in \Gamma_1^\mu$ and $b \in \Gamma_1^{m'}$, there is C such that for all $\varepsilon \in]0,1]$, all $\gamma \geq 0$ and all $u \in \mathrm{H}^{s+\mu+\mu'-1}$,*

$$\|\big(P_a^{\varepsilon,\gamma} P_b^{\varepsilon,\gamma} - P_{ab}^{\varepsilon,\gamma}\big)u\|_{s,\varepsilon,\gamma} \leq C\,\varepsilon\, \|u\|_{s+\mu+\mu'-1,\varepsilon,\gamma}\,.$$

Since the H_ε are unitary,

$$(P_a^{\varepsilon,\gamma})^* = (H_\varepsilon)^{-1}(T_{a^\varepsilon}^{\varepsilon\gamma})^* H_\varepsilon,$$

and Theorem (2.18) implies

PROPOSITION 2.23. *Consider a matrix valued symbol $a \in \Gamma_1^\mu$. There is C such that for all $\varepsilon \in]0,1]$, all $\gamma \geq 0$ and all $u \in \mathrm{H}^{s+\mu-1}$,*

$$\|((P_a^{\varepsilon,\gamma})^* - P_{a^*}^{\varepsilon,\gamma})u\|_{s,\varepsilon,\gamma} \leq C\varepsilon\, \|u\|_{s+\mu-1,\varepsilon,\gamma}\,.$$

Similarly, applying Theorem 2.19 to $T^{\gamma'}_{a^\varepsilon}$ and $T^{g'}_{w^\varepsilon}$ implies:

PROPOSITION 2.24. *Consider symbols $a \in \Gamma^\mu_1$ and $w \in \Gamma^0_1$. Suppose that there is $\chi \in \Gamma^0_1$ and $c > 0$ such that $\chi \geq 0$, $\chi w = w$ and*

$$\forall (x, \xi, \gamma) : \quad \chi^2(x, \xi, \gamma) \operatorname{Re} a(x, \xi, \gamma) \geq c\chi^2(x, \xi, \gamma)\, \Lambda^\mu(\xi, \gamma).$$

Then, there is C such that for all $\varepsilon \in]0, 1]$, all $\gamma \geq 0$ and all $u \in \mathrm{H}^{\mu/2}$

$$\frac{c}{2} \|P^{\varepsilon,\gamma}_w u\|^2_{\frac{\mu}{2},\varepsilon,\gamma} \leq \operatorname{Re}\left(P^{\varepsilon,\gamma}_a P^{\varepsilon,\gamma}_w u, P^{\varepsilon,\gamma}_w u\right)_{L^2} + C\varepsilon^2 \|u\|^2_{\frac{\mu}{2}-1,\varepsilon,\gamma} .$$

Finally, we note that for $a \in W^{1,\infty}(\mathbb{R}^d)$

$$au - P^{\varepsilon,\gamma}_a u = (H_\varepsilon)^{-1}\left(a^\varepsilon H_\varepsilon u - T^{\gamma'}_{a^\varepsilon} H_\varepsilon u\right),$$

with $\gamma' = \varepsilon\gamma$. Since $\partial_j H_\varepsilon u = \varepsilon H_\varepsilon \partial_j u$, one also has

$$a\partial_j u - P^{\varepsilon,\gamma}_a \partial_j u = \frac{1}{\varepsilon}(H_\varepsilon)^{-1}\left(a^\varepsilon \partial_j H_\varepsilon u - T^{\gamma'}_{a^\varepsilon} \partial_j H_\varepsilon u\right).$$

By Theorem 2.11, using that $\|\nabla a^\varepsilon\|_{L^\infty} = O(\varepsilon)$,

$$\|a^\varepsilon v - T^{\gamma'}_{a^\varepsilon} v\|_{1,\gamma'} \lesssim \varepsilon \|v\|_{0,\gamma'}$$
$$\gamma' \|a^\varepsilon v - T^{\gamma'}_{a^\varepsilon} v\|_{L^2} \lesssim \varepsilon \|u\|_{\frac{p}{p_0}-1,\gamma'}$$
$$\|a^\varepsilon \partial_j v - T^{\gamma'}_{a^\varepsilon} \partial_j v\|_{L^2} \lesssim \varepsilon \|v\|_{\frac{p}{p_j}-1,\gamma'} .$$

Using (2.49), this implies

PROPOSITION 2.25. *For all $a \in W^{1,\infty}$, there is C such that for all $\varepsilon \in]0, 1]$, all $\gamma \geq 0$ and all $u \in L^2$,*

$$\|au - P^{\varepsilon,\gamma}_a u\|_{1,\varepsilon,\gamma} \leq C\varepsilon \|u\|_{0,\varepsilon,\gamma} ,$$
$$\gamma \|au - P^{\varepsilon,\gamma}_a u\|_{L^2} \leq C \|u\|_{\frac{p}{p_0}-1,\varepsilon,\gamma} ,$$
$$\|a\partial_j u - P^{\varepsilon,\gamma}_a \partial_j u\|_{L^2} \leq C \|v\|_{\frac{p}{p_j}-1,\varepsilon,\gamma} .$$

2.3. The homogeneous calculus. We now prove the results announced in section 3.1 and we take the notations used in that section. On one hand, we consider the variables $\widetilde{y} = (t, y) \in \mathbb{R}^d$. On the other hand we consider and extra variable $x \in \mathbb{R}$, considered as a parameter.

Denote by $\widetilde{\eta}$ the dual variable of $\widetilde{y}$. We consider here the homogeneous case where all the weights p_j and p in (2.1) are equal to one. In this case $\langle\zeta\rangle = |\zeta| = (\gamma^2 + \sum(\widetilde{y}_j)^2)^{1/2}$. We restrict attention to $\gamma \geq 1$, and thus $\Lambda(\zeta) \approx |\zeta|$. In this framework, the Definition 3.1 of class of symbols Γ^μ_k coincide with Definition 2.4. For $a \in \Gamma^\mu_0$, the operator T^γ_a is thus defined accordingly. In the homogeneous case, the spaces $\mathrm{H}^s(\mathbb{R}^d)$ introduced in section B1, are the usual Sobolev spaces $H^s(\mathbb{R}^d)$ and the norms in (2.12) are equivalent to the norms (3.5), with constant independent of $\gamma \geq 1$.

A symbol a is in the class $\Gamma^\mu_{k,0}$ introduced in Definition 3.2, if and only if $\{a(x, \cdot) : x \in \mathbb{R}\}$ is a bounded family in Γ^μ_k. Thus the definition

$$(T^\gamma_a u)(x, \cdot) = T^\gamma_{a(x,\cdot)} u(x, \cdot)$$

makes sense for u in the space $L^2(\mathbb{R}; H^s(\mathbb{R}^d))$, called $\mathcal{H}^{0,s}$ in section 3.1.

As in section 3, introduce the vector fields $Z_0 = x\partial_x$ and for $j \in \{1, \dots, d\}$, $Z_j = \partial_{\widetilde{y}_j}$. For $a \in \Gamma^\mu_{0,0}$ such that $Z_j a \in \Gamma^\mu_{0,0}$, the definition of T^γ shows that

$$Z_j T^\gamma_a u = T^\gamma_a Z_j u + T^\gamma_{Z_j a} u \,.$$

Indeed, for the pseudodifferential operators, $\mathrm{Op}^\gamma(\sigma)$ defined in (2.24), one has $[\partial_{\widetilde{y}_j}, \mathrm{Op}^\gamma(\sigma)] = \mathrm{Op}^\gamma(\partial_{\widetilde{y}_j}\sigma)$. Moreover, x being a parameter, one also has $[x\partial_x, \mathrm{Op}^\gamma(\sigma)] = \mathrm{Op}^\gamma(x\partial_x\sigma)$. Finally, $T^\gamma_a = \mathrm{Op}^\gamma(\sigma_a)$ and $Z_j\sigma_a = \sigma_{Z_j a}$ since the mapping $a \mapsto \sigma_a$ is given by a convolution in the variables $\widetilde{y}$, x and $(\widetilde{\eta}, \gamma)$ being parameters. More generally, the following Leibniz' formula is valid for $a \in \Gamma^\mu_{0,m}$ and $|\alpha| \le m$:

$$Z^\alpha T^\gamma_a = \sum_{\beta \le \alpha} \binom{\alpha}{\beta} T^\gamma_{Z^\beta a}\, Z^{\alpha - \beta} \,. \tag{2.50}$$

Proposition 2.9 implies, with notations as in section 3.1, that for $a \in \Gamma^\mu_{0,0}$, there is C such that for all $x \in \mathbb{R}$ and all $\gamma \ge 1$:

$$\big|(T^\gamma_a u)(x, \cdot)\big|_{0,s,\gamma} \le C\, \big|u(x, \cdot)\big|_{0,s+\mu,\gamma}.$$

Squaring and integrating in x, yields

$$\|T^\gamma_a u\|_{0,s,\gamma} \le C\, \|u\|_{0,s+\mu,\gamma} \,.$$

Using (2.50), one obtains that for $a \in \Gamma^\mu_{0,m}$ one has, for $\gamma \ge 1$,

$$\|T^\gamma_a u\|_{m,s,\gamma} \le C\, \|u\|_{m,s+\mu,\gamma} \,.$$

This proves Proposition 3.4.

Similarly, Theorem(2.16) implies that for $a \in \Gamma^\mu_{1,0}$ and $b \in \Gamma^{\mu'}_{1,0}$, $R^\gamma(a,b) = T^\gamma_a \circ T^\gamma_b - T^\gamma_{ab}$ satisfies

$$\big|(R^\gamma(a,b)u)(x, \cdot)\big|_{0,s,\gamma} \le C\, \big|u(x, \cdot)\big|_{0,s+\mu+\mu'-1,\gamma}$$

and thus

$$\|(R^\gamma(a,b)u)\|_{0,s,\gamma} \le C\, \|u\|_{0,s+\mu+\mu'-1,\gamma}$$

where C is independent of $\gamma \ge 1$ and u. Repeated applications of Leibniz' rule (2.50) implies that

$$Z^\alpha R^\gamma(a,b) = \sum_{\alpha^1+\alpha^2+\alpha^3=\alpha} \frac{\alpha!}{\alpha^1!\,\alpha^2!\,\alpha^3!} R^\gamma(Z^{\alpha^1} a,\, Z^{\alpha^2} b)\, Z^{\alpha^3} \,.$$

Therefore, for $a \in \Gamma^\mu_{1,m}$ and $b \in \Gamma^{\mu'}_{1,m}$, there is C such that for all $\gamma \ge 1$ and all $u \in \mathcal{H}^{m,s+\mu+\mu'-1}$,

$$\|(R^\gamma(a,b)u)\|_{m,s,\gamma} \le C\, \|u\|_{m,s+\mu+\mu'-1,\gamma}$$

and Proposition 3.5 is proved.

Propositions 3.6 and 3.7, which involve estimates with $m = 0$, directly follow from Theorems 2.18 and 2.19.

Next, Theorem 2.11 implies that for $a \in L^\infty$ such that $\nabla_{\widetilde{y}} a \in L^\infty$, the operators $R^\gamma_a u = \gamma(au - T^\gamma_a u)$ and $R^\gamma_a u = a\partial_{\widetilde{y}_j} u - T^\gamma_a \partial_{\widetilde{y}_j} u$, satisfy

$$\|(R^\gamma_a u)\|_{0,1,\gamma} \le C \|\nabla_{\widetilde{y}} a\|_{L^\infty} \|u\|_{0,0,\gamma}.$$

Since

$$Z^\alpha R^\gamma_a = \sum_{\beta \le \alpha} \binom{\alpha}{\beta} R^\gamma_{Z^\beta a}\, Z^{\alpha-\beta} \,,$$

this implies that for a in the space $\mathcal{W}^{m,1}$, and $u \in \mathcal{H}^{m,O}$,

$$\|(R_a^\gamma u)\|_{m,1,\gamma} \leq C\|a\|_{\mathcal{W}^{m,1}}\|u\|_{m,0,\gamma} ,$$

implying Proposition 3.9.

2.4. The semi-classical parabolic calculus. We now prove the results announced in section 3.2. We consider the parabolic weights $p_0 = p_1 = 1$ for the variables γ and τ, and $p_j = 2$ for the spatial tangential variables y_j (compare (3.11) and (2.1)). $x \in \mathbb{R}$ is an extra variable.

The class of symbols $\mathrm{P}\Gamma_k^\mu$ in Definition 3.10 is the class Γ_k^μ of Definition 2.4 corresponding to our present quasi-homogeneous structure. For a is in the class $\mathrm{P}\Gamma_{k,0}^\mu$, $\{a(x,\cdot) : x \in \mathbb{R}\}$ is a bounded family in $\mathrm{P}\Gamma_k^\mu$ and thus we can define

$$(P_a^{\varepsilon,\gamma} u)(x,\cdot) = P_{a(x,\cdot)}^{\varepsilon,\gamma} u(x,\cdot)$$

for u in the space $L^2(\mathbb{R}; \mathrm{H}^s(\mathbb{R}^d))$, called $P\mathcal{H}^{0,s}$ in section 3.2.

We consider again the vector fields $Z_0 = x\partial_x$ and $Z_j = \partial_{\widetilde{y}_j}$. For $a \in \mathrm{P}\Gamma_{1,0}^\mu$, one has

$$Z_j P_a^{\varepsilon,\gamma} u = P_a^{\varepsilon,\gamma} Z_j u + P_{Z_j a}^{\varepsilon,\gamma} u . \tag{2.51}$$

To prove this identity, we use the definition (2.47) of $P^{\varepsilon,\gamma}$. The equality (2.51) is quite clear for Z_0 since H_ε and $x\partial_x$ commute and $[x\partial_x, T_{a^\varepsilon}^{\gamma'}] = T_{x\partial_x a^\varepsilon}^{\gamma'} = T_{(x\partial_x a)^\varepsilon}^{\gamma'}$. Next, for $j \in \{i, \ldots, d\}$, (2.51) follows directly from the identities :

$$\begin{aligned} &\partial_{\widetilde{y}_j}(H_\varepsilon)^{-1} = \varepsilon^{-1}(H_\varepsilon)^{-1}\partial_{\widetilde{y}_j} , \\ &[\partial_{\widetilde{y}_j}, T_{a^\varepsilon}^{\gamma'}] = T_{\partial_{\widetilde{y}_j}(a^\varepsilon)}^{\gamma'} = \varepsilon T_{(\partial_{\widetilde{y}_j} a)^\varepsilon}^{\gamma'} , \\ &\partial_{\widetilde{y}_j} H_\varepsilon = \varepsilon H_\varepsilon \partial_{\widetilde{y}_j} . \end{aligned}$$

The identity (2.51) extends to $a \in \mathrm{P}\Gamma_{0,m}^\mu$ and $|\alpha| \leq m$:

$$Z^\alpha P_a^{\varepsilon,\gamma} = \sum_{\beta \leq \alpha} \binom{\alpha}{\beta} P_{Z^\beta a}^{\varepsilon,\gamma} Z^{\alpha-\beta} . \tag{2.52}$$

As in section B.3, the Propositions of section B.2 immediately imply Propositions 3.14 (adjoints) and 3.15 (Garding's inequality) and the estimates in Propositions 3.12 and 3.13 when $m = 0$. Using Leibniz's formula and (2.52) they immediately extends to general m, as in section B.3. The localization of the spectrum of $P_a^{\varepsilon,\gamma} u$ when a is compactly supported or supported outside a ball, as stated in *iii)* and *iv)* of Proposition 3.12, follows directly from Remark 2.10.

Moreover, Proposition 2.25 immediately implies

$$\|au - P_a^{\varepsilon,\gamma} u\|_{0,1,\varepsilon,\gamma} + \sum_{|\alpha|=1} \varepsilon\|a\partial_y^\alpha u - P_a^{\varepsilon,\gamma}\partial_y^\alpha u\|_{0,0,\varepsilon,\gamma} \lesssim \varepsilon\|u\|_{0,0,\varepsilon,\gamma}$$

and

$$\gamma\|au - P_a^{\varepsilon,\gamma} u\|_{0,0,\varepsilon,\gamma} + \|a\partial_t u - P_a^{\varepsilon,\gamma}\partial_t u\|_{0,0,\varepsilon,\gamma} \lesssim \|u\|_{0,1,\varepsilon,\gamma} .$$

Moreover, it also implies that

$$\sum_{|\alpha|=2} \varepsilon\|a\partial_y^\alpha u - P_a^{\varepsilon,\gamma}\partial_y^\alpha u\|_{0,0,\varepsilon,\gamma} \lesssim \|\varepsilon\nabla_y u\|_{0,0,\varepsilon,\gamma} \lesssim \|u\|_{0,1,\varepsilon,\gamma} .$$

This shows that the estimates (3.22) and (3.23) of Proposition 3.17 are satisfied when $m = 0$. Commuting with the Z^α and using the Leibniz' formula (2.51) to the remainder $R_a^{\varepsilon,\gamma}u = au - P_a^{\varepsilon,\gamma}u$, we obtain the estimates for $m > 0$ and the Proposition 3.17 follows.

To prove Corollary 3.18, we use that

$$\|R\|_{m,2,\varepsilon,\gamma} \lesssim \|R\|_{m,0,\varepsilon,\gamma} + \varepsilon\gamma\|R\|_{m,0,\varepsilon,\gamma} + \varepsilon\|\partial_t R\|_{m,0,\varepsilon,\gamma} + \varepsilon\|\nabla_y R\|_{m,0,\varepsilon,\gamma} + \varepsilon^2\|\nabla_y^2 R\|_{m,0,\varepsilon,\gamma}\,.$$

We apply this estimate to $R = R(a,u) := au - P_a^{\varepsilon,\gamma}u$ and show that each term is bounded by

$$C(\varepsilon\|\nabla a\|_{L^\infty} + \varepsilon^2\|\nabla^2 a\|_{L^\infty})\|u\|_{m,1,\varepsilon,\gamma}$$

This is true for the first two terms, by Proposition 3.17. Next we write that $\partial R(a,u) = R(a,\partial u) + R(\partial a, u)$ and $\partial_y^2 R(a,u)$ as a linear combination of $R(a,\partial_y^2 u)$, $R(\partial_y a, \partial_y u)$ and $R(\partial_y^2 a, u)$. The terms $R(b,u)$, $b = \partial_{t,y} a$ or $b = \partial_y^2 a$ are controlled by Proposition 3.12 for the part $P_b^{\varepsilon,\gamma}u$ and directly by Leibniz's rule for the part bu. The other terms are estimated by Proposition 3.17 applied to u or $\varepsilon\partial_j u$.

It only remains to prove Proposition 3.19.

Proof of Proposition 3.19. The goal is to compare $P_b^{\varepsilon,\gamma}$ and $T_{\mathrm{b}^\varepsilon}^\gamma$ when $b \in \mathbb{P}\Gamma_{1,0}^0$ has compact support in $\zeta = (\widetilde{\eta}, \gamma)$ and $\mathrm{b}^\varepsilon(x,\widetilde{y},\zeta) = b(x,\widetilde{y},\varepsilon\zeta)$.

a) First, we note that b can be expanded into a rapidly convergent series

$$b(x,\widetilde{y},\zeta) = \sum_\nu b_\nu(x,\widetilde{y})\, c_\nu(\zeta),$$

where $c_\nu \in C_0^\infty(\mathbb{R}^d)$, $b_\nu \in W^{1,\infty}(\mathbb{R}^{1+d})$ and for all $\alpha \in \mathbb{N}^{d+1}$, the sequence $\|b_\nu\|_{W^{1,\infty}}\|\partial^\alpha c_\nu\|_{L^\infty}$ is rapidly decreasing. Indeed, since b has compact support in ζ, there is $\theta \in C_0^\infty$ such that $\theta(\zeta)b = b$. Let L be so large that the support of θ is contained in the box $\mathbb{B} = [-L, L]^{d+1}$. Then, since b is C^∞, $\theta(\cdot)\, b(x,\widetilde{y},\cdot)$ can be expanded into rapidly convergent Fourier series, yielding

$$b(x,\widetilde{y},\zeta) = \sum_{\nu\in\mathbb{Z}^{d+1}} b_\nu(x,\widetilde{y})\,\theta(\zeta)e^{2i\pi\nu\zeta/L}\,.$$

b) Next, we note that when $c = c(\zeta)$ and $\mathrm{c}^\varepsilon(\zeta) = c(\varepsilon\zeta)$, then

$$P_c^{\varepsilon,\gamma} = T_{\mathrm{c}^\varepsilon}^\gamma = c(\varepsilon D_{\widetilde{y}}, \varepsilon\gamma)$$

is the operator defined by the Fourier multiplier $\widetilde{\eta} \mapsto c(\varepsilon\widetilde{\eta}, \varepsilon\gamma)$. Moreover, the definitions imply that

$$P_{b_\nu c_\nu}^{\varepsilon,\gamma} = P_{b_\nu}^{\varepsilon,\gamma} P_{c_\nu}^{\varepsilon,\gamma}\,, \qquad T_{b_\nu \mathrm{c}_\nu^\varepsilon}^\gamma = T_{b_\nu}^\gamma T_{\mathrm{c}_\nu^\varepsilon}^\gamma\,.$$

Therefore

$$P_b^{\varepsilon,\gamma} - T_{\mathrm{b}^\varepsilon}^\gamma = \sum_\nu \big(P_{b_\nu}^{\varepsilon,\gamma} - T_{b_\nu}^\gamma\big) P_{c_\nu}^{\varepsilon,\gamma}\,.$$

Since $\partial_{\widetilde{y}_j}$ and $P_{c_\nu}^{\varepsilon,\gamma}$ commute, one also has

$$(P_b^{\varepsilon,\gamma} - T_{\mathrm{b}^\varepsilon}^\gamma)\partial_{\widetilde{y}_j} = \sum_\nu \big(P_{b_\nu}^{\varepsilon,\gamma}\partial_{\widetilde{y}_j} - T_{b_\nu}^\gamma\partial_{\widetilde{y}_j}\big) P_{c_\nu}^{\varepsilon,\gamma}\,.$$

Propositions 3.9 and 3.17 imply that

$$\gamma\|b_\nu u - T_{b_\nu}^\gamma u\|_{L^2} + \|b_\nu\nabla_{\widetilde{y}} u - T_{b_\nu}^\gamma \nabla_{\widetilde{y}} u\|_{L^2} \lesssim \|\nabla_{\widetilde{y}} b_\nu\|_{L^\infty}\|u\|_{L^2}\,,$$

$$\gamma\|b_\nu u - P_{b_\nu}^{\varepsilon,\gamma} u\|_{L^2} + \|b_\nu\nabla_{\widetilde{y}} u - P_{b_\nu}^{\varepsilon,\gamma}\nabla_{\widetilde{y}} u\|_{L^2} \lesssim \|\nabla_{\widetilde{y}} b_\nu\|_{L^\infty}\|u\|_{0,1,\varepsilon,\gamma}\,.$$

Moreover,

$$\|T_{\mathrm{c}_\nu^\varepsilon}^\gamma u\|_{L^2} \le \|c_\nu\|_{L^\infty}\,\|u\|_{L^2}\,,$$

and, since c_ν is supported in a fixed compact set,

$$\|P^{\varepsilon,\gamma}_{c_\nu} u\|_{0,1,\varepsilon,\gamma} \lesssim \|c_\nu\|_{L^\infty} \|u\|_{L^2}.$$

Therefore,

$$\gamma\|T^\gamma_{b_\nu c^\varepsilon_\nu} - P^{\varepsilon,\gamma}_{b_\nu c_\nu} u\|_{L^2} + \|T^\gamma_{b_\nu c^\varepsilon_\nu} \nabla_{\tilde{y}} u - P^{\varepsilon,\gamma}_{b_\nu c_\nu} \nabla_{\tilde{y}} u\|_{L^2} \lesssim \|\nabla_{\tilde{y}} b_\nu\|_{L^\infty} \|c_\nu\|_{L^\infty} \|u\|_{L^2}\,.$$

Since the series $\sum \|\nabla_{\tilde{y}} b_\nu\|_{L^\infty} \|c_\nu\|_{L^\infty}$ is convergent, we conclude that

$$\gamma\|T^\gamma_{b^\varepsilon} - P^{\varepsilon,\gamma}_b u\|_{L^2} + \|T^\gamma_{b^\varepsilon} \nabla_{\tilde{y}} u - P^{\varepsilon,\gamma}_b \nabla_{\tilde{y}} u\|_{L^2 s} \lesssim \|u\|_{L^2}\,. \tag{2.53}$$

c) For all temperate distribution u and $\gamma \geq 1$, there holds

$$u = \gamma v_0 + \sum \partial_{\tilde{y}_j} v_j,$$

where the Fourier transform of the v_j' are given by

$$\widehat{v}_0(\tilde{\eta}) = \gamma|\zeta|^{-2}\widehat{u}(\tilde{\eta})\,, \quad \widehat{v}_j(\tilde{\eta}) = -i\tilde{\eta}_j/|\zeta|^{-2}\widehat{u}(\tilde{\eta})\,.$$

Thus, for all j, one has in the scale of norms (3.6):

$$\|v_j\|_{L^2} \leq \|u\|_{0,-1,\gamma}.$$

Therefore, using (2.53) for the v_j, we obtain that

$$\|T^\gamma_{b^\varepsilon} u - P^{\varepsilon,\gamma}_b u\|_{0,0,\gamma} \lesssim \|u\|_{0,-1,\gamma}\,,$$

finishing the proof of Proposition 3.19. □

Bibliography

[Ag] S. Agmon, *Problèmes mixtes pour les équations hyperboliques d'ordre supérieur, Les Équations aux Dérivées Partielles.* Éditions du Centre National de la Recherche Scientifique, Paris (1962) 13–18.

[AGJ] J. Alexander, R. Gardner, and C.K.R.T. Jones, *A topological invariant arising in the analysis of traveling waves.* J. Reine Angew. Math. 410 (1990) 167–212.

[AMPZ] A. Azevedo, D. Marchesin, B. Plohr, and K. Zumbrun, *Nonuniqueness of solutions of Riemann problems.* Z. Angew. Math. Phys. 47 (1996), 977–998.

[BBB] C.Bardos, D.Brezis, and H.Brezis, *Perturbations singulières et prolongement maximaux d'opérateurs positifs*, Arch. Rational Mech. Anal., 53 (1973), 69–100.

[Ba-Ra] C.Bardos and J.Rauch, *Maximal positive boundary value problems as limits of singular perturbation problems*, Trans. Amer. Math. Soc., 270 (1982), 377–408.

[Bo] J.M. Bony, *Calcul symbolique et propagation des singularités pour les équations aux dérivées partielles non linéaires*, Ann. Sc. E.N.S. Paris, 14 (1981) pp 209-246.

[Ch-P] J. Chazarain and A. Piriou, *Introduction to the theory of linear partial differential equations,* Translated from the French. Studies in Mathematics and its Applications, 14. North-Holland Publishing Co., Amsterdam-New York, 1982. xiv+559 pp. ISBN: 0-444-86452-0.

[Co-Me] R.Coifmann and Y.Meyer, *Au delà des opérateurs pseudo-différentiels*, Astérisque 57, (1978).

[E1] J.W. Evans, *Nerve axon equations: I. Linear approximations.* Ind. Univ. Math. J. 21 (1972) 877–885.

[E2] ———, *Nerve axon equations: II. Stability at rest.* Ind. Univ. Math. J. 22 (1972) 75–90.

[E3] ———, *Nerve axon equations: III. Stability of the nerve impulse.* Ind. Univ. Math. J. 22 (1972) 577–593.

[E4] ———, *Nerve axon equations: IV. The stable and the unstable impulse.* Ind. Univ. Math. J. 24 (1975) 1169–1190.

[GZ] R. Gardner and K. Zumbrun, *The Gap Lemma and geometric criteria for instability of viscous shock profiles.* Comm. Pure Appl. Math. 51 (1998), 797–855.

[Gi-Se] M. Gisclon and D. Serre, *Conditions aux limites pour un système strictement hyperbolique fournies par le schéma de Godunov.* RAIRO Modél. Math. Anal. Numér. 31 (1997), 359–380.

[Gr-Gu] E. Grenier and O. Gues, *Boundary layers for viscous perturbations of noncharacteristic quasilinear hyperbolic problems.* J. Differential Equations 143 (1998), 110–146.

[Gr-Ro] E. Grenier and F. Rousset, *Stability of one-dimensional boundary layers by using Green's functions.* Comm. Pure Appl. Math. 54 (2001), 1343–1385.

[Gu1] O. Guès, *Perturbations visqueuses de problèmes mixtes hyperboliques et couches limites*, Ann.Inst.Fourier, 45 (1995),973–1006.

[Gu2] ———, *Problème mixte hyperbolique quasilinéaire caractéristique*, Comm. in Part.Diff.Equ., 15 (1990), 595–645.

[HoZ] D. Hoff and K. Zumbrun, *Multi-dimensional diffusion waves for the Navier-Stokes equations of compressible flow.* Indiana Univ. Math. J. 44 (1995), 603–676.

[Hör] L. Hörmander, *Lectures on Nonlinear Hyprbolic Differential Equations*, Mathématiques et Applications 26, Sringer Verlag, 1997.

[J] C.K.R.T. Jones, *Stability of the travelling wave solution of the FitzHugh–Nagumo system.* Trans. Amer. Math. Soc. 286 (1984), 431–469.

[K] T. Kapitula, *On the stability of travelling waves in weighted L^∞ spaces.* J. Diff. Eqs. 112 (1994), 179–215.

[Kr] H.O. Kreiss, *Initial boundary value problems for hyperbolic systems,* Comm. Pure Appl. Math. 23 (1970) 277-298.

[Lio] J.L. Lions, *Perturbations singulières dans les problèmes aux limites et en contrôle optimal,* Lectures Notes in Math., 323, Sringer Verlag, 1973.

[Maj] A. Majda, *The stability of multi-dimensional shock fronts – a new problem for linear hyperbolic equations.* Mem. Amer. Math. Soc. 275 (1983).

[Mé1] G. Métivier, *Interaction de deux chocs pour un système de deux lois de conservation, en dimension deux d'espace.* Trans. Amer. Math. Soc. 296 (1986) 431–479.

[Mé2] ———, *Stability of multidimensional shocks.* Advances in the theory of shock waves, 25–103, Progr. Nonlinear Differential Equations Appl., 47, Birkhäuser Boston, Boston, MA, 2001.

[Mé3] ———, *The Block Structure Condition for Symmetric Hyperbolic Problems,* Bull. London Math. Soc., 32 (2000), 689–702

[Mey] Y. Meyer, *Remarques sur un théorème de J.M. Bony,* Supplemento al Rendiconti der Circolo Matematico di Palermo, Serie II, No. 1, 1981.

[Mok] A. Mokrane, *Problèmes mixtes hyperboliques non linéaires,* Thesis, Université de Rennes 1, 1987.

[PW] R. L. Pego and M.I. Weinstein, *Eigenvalues, and instabilities of solitary waves.* Philos. Trans. Roy. Soc. London Ser. A 340 (1992), 47–94.

[Ral] J.V. Ralston, *Note on a paper of Kreiss,* Comm. Pure Appl. Math. 24 (1971) 759–762.

[Ra1] J. Rauch, *BV estimates fail for most quasilinear hyperbolic systems in dimensions greater than one,* Comm. Math. Phys., 106 (1986), 481–484.

[Ra2] ———, *Symmetric positive systems with boundary characteristic of constant multiplicity,* Trans. Amer. Math. Soc., 291 (1985), 167–185.

[Ra-Ma] J. Rauch and F. Massey, *Differentiability of solutions to hyperbolic intitial boundary value problems.* Trans. Amer. Math. Soc. 189 (1974) 303-318.

[Rou] F. Rousset, *Inviscid boundary conditions and stability of viscous boundary layers.* Asymptot. Anal. 26 (2001), no. 3-4, 285–306.

[S] D. Serre, *Sur la stabilité des couches limites de viscosité.* Ann. Inst. Fourier (Grenoble) 51 (2001), 109–130.

[St] E. Stein, *Singular Integrals and Differentiability Propseties of Functions,* Princeton Univ. Press, Princeton, 1970.

[Tay] M. Taylor, *Partial Differential Equations* III, Applied Mathematical Sciences 117, Springer, 1996.

[ZS] K. Zumbrun and D.Serre, *Viscous and inviscid stability of multidimensional planar shock fronts.* Indiana Univ. Math. J. 48 (1999), 937–992.

[ZH] K. Zumbrun and P. Howard, *Pointwise semigroup methods and stability of viscous shock waves.* Indiana Mathematics Journal V47 (1998), 741–871.

[Z] K. Zumbrun, *Multidimensional stability of planar viscous shock waves.* Advances in the theory of shock waves, 307–516, Progr. Nonlinear Differential Equations Appl., 47, Birkhäuser Boston, Boston, MA, 2001.

Editorial Information

To be published in the *Memoirs*, a paper must be correct, new, nontrivial, and significant. Further, it must be well written and of interest to a substantial number of mathematicians. Piecemeal results, such as an inconclusive step toward an unproved major theorem or a minor variation on a known result, are in general not acceptable for publication. Papers appearing in *Memoirs* are generally longer than those appearing in *Transactions*, which shares the same editorial committee.

As of January 31, 2005, the backlog for this journal was approximately 5 volumes. This estimate is the result of dividing the number of manuscripts for this journal in the Providence office that have not yet gone to the printer on the above date by the average number of monographs per volume over the previous twelve months, reduced by the number of volumes published in four months (the time necessary for preparing a volume for the printer). (There are 6 volumes per year, each containing at least 4 numbers.)

A Consent to Publish and Copyright Agreement is required before a paper will be published in the *Memoirs*. After a paper is accepted for publication, the Providence office will send a Consent to Publish and Copyright Agreement to all authors of the paper. By submitting a paper to the *Memoirs*, authors certify that the results have not been submitted to nor are they under consideration for publication by another journal, conference proceedings, or similar publication.

Information for Authors

Memoirs are printed from camera copy fully prepared by the author. This means that the finished book will look exactly like the copy submitted.

The paper must contain a *descriptive title* and an *abstract* that summarizes the article in language suitable for workers in the general field (algebra, analysis, etc.). The *descriptive title* should be short, but informative; useless or vague phrases such as "some remarks about" or "concerning" should be avoided. The *abstract* should be at least one complete sentence, and at most 300 words. Included with the footnotes to the paper should be the 2000 *Mathematics Subject Classification* representing the primary and secondary subjects of the article. The classifications are accessible from `www.ams.org/msc/`. The list of classifications is also available in print starting with the 1999 annual index of *Mathematical Reviews*. The Mathematics Subject Classification footnote may be followed by a list of *key words and phrases* describing the subject matter of the article and taken from it. Journal abbreviations used in bibliographies are listed in the latest *Mathematical Reviews* annual index. The series abbreviations are also accessible from `www.ams.org/publications/`. To help in preparing and verifying references, the AMS offers MR Lookup, a Reference Tool for Linking, at `www.ams.org/mrlookup/`. When the manuscript is submitted, authors should supply the editor with electronic addresses if available. These will be printed after the postal address at the end of the article.

Electronically prepared manuscripts. The AMS encourages electronically prepared manuscripts, with a strong preference for $\mathcal{A}\!\mathcal{M}\!\mathcal{S}$-LaTeX. To this end, the Society has prepared $\mathcal{A}\!\mathcal{M}\!\mathcal{S}$-LaTeX author packages for each AMS publication. Author packages include instructions for preparing electronic manuscripts, the *AMS Author Handbook*, samples, and a style file that generates the particular design specifications of that publication series. Though $\mathcal{A}\!\mathcal{M}\!\mathcal{S}$-LaTeX is the highly preferred format of TeX, author packages are also available in $\mathcal{A}\!\mathcal{M}\!\mathcal{S}$-TeX.

Authors may retrieve an author package from e-MATH starting from `www.ams.org/tex/` or via FTP to `ftp.ams.org` (login as `anonymous`, enter username as password, and type `cd pub/author-info`). The *AMS Author Handbook* and the *Instruction Manual* are available in PDF format following the author packages link from `www.ams.org/tex/`. The author package can be obtained free of charge by sending email to `pub@ams.org` (Internet) or from the Publication Division, American Mathematical Society, 201 Charles St., Providence, RI 02904, USA. When requesting an author package, please specify $\mathcal{A}\mathcal{M}\mathcal{S}$-LaTeX or $\mathcal{A}\mathcal{M}\mathcal{S}$-TeX, Macintosh or IBM (3.5) format, and the publication in which your paper will appear. Please be sure to include your complete mailing address.

Sending electronic files. After acceptance, the source file(s) should be sent to the Providence office (this includes any TeX source file, any graphics files, and the DVI or PostScript file).

Before sending the source file, be sure you have proofread your paper carefully. The files you send must be the EXACT files used to generate the proof copy that was accepted for publication. For all publications, authors are required to send a printed copy of their paper, which exactly matches the copy approved for publication, along with any graphics that will appear in the paper.

TeX files may be submitted by email, FTP, or on diskette. The DVI file(s) and PostScript files should be submitted only by FTP or on diskette unless they are encoded properly to submit through email. (DVI files are binary and PostScript files tend to be very large.)

Electronically prepared manuscripts can be sent via email to `pub-submit@ams.org` (Internet). The subject line of the message should include the publication code to identify it as a Memoir. TeX source files, DVI files, and PostScript files can be transferred over the Internet by FTP to the Internet node `e-math.ams.org` (130.44.1.100).

Electronic graphics. Comprehensive instructions on preparing graphics are available at `www.ams.org/jourhtml/graphics.html`. A few of the major requirements are given here.

Submit files for graphics as EPS (Encapsulated PostScript) files. This includes graphics originated via a graphics application as well as scanned photographs or other computer-generated images. If this is not possible, TIFF files are acceptable as long as they can be opened in Adobe Photoshop or Illustrator. No matter what method was used to produce the graphic, it is necessary to provide a paper copy to the AMS.

Authors using graphics packages for the creation of electronic art should also avoid the use of any lines thinner than 0.5 points in width. Many graphics packages allow the user to specify a "hairline" for a very thin line. Hairlines often look acceptable when proofed on a typical laser printer. However, when produced on a high-resolution laser imagesetter, hairlines become nearly invisible and will be lost entirely in the final printing process.

Screens should be set to values between 15% and 85%. Screens which fall outside of this range are too light or too dark to print correctly. Variations of screens within a graphic should be no less than 10%.

Inquiries. Any inquiries concerning a paper that has been accepted for publication should be sent directly to the Electronic Prepress Department, American Mathematical Society, 201 Charles St., Providence, RI 02904, USA.

Editors

This journal is designed particularly for long research papers, normally at least 80 pages in length, and groups of cognate papers in pure and applied mathematics. Papers intended for publication in the *Memoirs* should be addressed to one of the following editors. In principle the Memoirs welcomes electronic submissions, and some of the editors, those whose names appear below with an asterisk (*), have indicated that they prefer them. However, editors reserve the right to request hard copies after papers have been submitted electronically. Authors are advised to make preliminary email inquiries to editors about whether they are likely to be able to handle submissions in a particular electronic form.

***Algebra** to ALEXANDER KLESHCHEV, Department of Mathematics, University of Oregon, Eugene, OR 97403-1222; email: ams@noether.uoregon.edu

Algebraic geometry to DAN ABRAMOVICH, Department of Mathematics, Brown University, Box 1917, Providence, RI 02912; email: amsedit@math.brown.edu

***Algebraic number theory** to V. KUMAR MURTY, Department of Mathematics, University of Toronto, 100 St. George Street, Toronto, ON M5S 1A1, Canada; email: murty@math.toronto.edu

***Algebraic topology** to ALEJANDRO ADEM, Department of Mathematics, University of British Columbia, Room 121, 1984 Mathematics Road, Vancouver, British Columbia, Canada V6T 1Z2; email: adem@math.ubc.ca

Combinatorics and Lie theory to SERGEY FOMIN, Department of Mathematics, University of Michigan, Ann Arbor, Michigan 48109-1109; email: fomin@umich.edu

Complex analysis and harmonic analysis to ALEXANDER NAGEL, Department of Mathematics, University of Wisconsin, 480 Lincoln Drive, Madison, WI 53706-1313; email: nagel@math.wisc.edu

***Differential geometry and global analysis** to LISA C. JEFFREY, Department of Mathematics, University of Toronto, 100 St. George St., Toronto, ON Canada M5S 3G3; email: jeffrey@math.toronto.edu

Dynamical systems and ergodic theory to ROBERT F. WILLIAMS, Department of Mathematics, University of Texas, Austin, Texas 78712-1082; email: bob@math.utexas.edu

***Functional analysis and operator algebras** to MARIUS DADARLAT, Department of Mathematics, Purdue University, 150 N. University St., West Lafayette, IN 47907-2067; email: mdd@math.purdue.edu

***Geometric analysis** to TOBIAS COLDING, Courant Institute, New York University, 251 Mercer St., New York, NY 10012; email: traneditor@cims.nyu.edu

***Geometric analysis** to MLADEN BESTVINA, Department of Mathematics, University of Utah, 155 South 1400 East, JWB 233, Salt Lake City, Utah 84112-0090; email: bestvina@math.utah.edu

Harmonic analysis, representation theory, and Lie theory to ROBERT J. STANTON, Department of Mathematics, The Ohio State University, 231 West 18th Avenue, Columbus, OH 43210-1174; email: stanton@math.ohio-state.edu

***Logic** to STEFFEN LEMPP, Department of Mathematics, University of Wisconsin, 480 Lincoln Drive, Madison, Wisconsin 53706-1388; email: lempp@math.wisc.edu

Number theory to HAROLD G. DIAMOND, Department of Mathematics, University of Illinois, 1409 W. Green St., Urbana, IL 61801-2917; email: diamond@math.uiuc.edu

***Ordinary differential equations, and applied mathematics** to PETER W. BATES, Department of Mathematics, Michigan State University, East Lansing, MI 48824-1027; email: bates@math.msu.edu

***Partial differential equations** to PATRICIA E. BAUMAN, Department of Mathematics, Purdue University, West Lafayette, IN 47907-1395; email: bauman@math.purdue.edu

***Probability and statistics** to KRZYSZTOF BURDZY, Department of Mathematics, University of Washington, Box 354350, Seattle, Washington 98195-4350; email: burdzy@math.washington.edu

***Real analysis and partial differential equations** to DANIEL TATARU, Department of Mathematics, University of California, Berkeley, Berkeley, CA 94720; email: tataru@math.berkeley.edu

All other communications to the editors should be addressed to the Managing Editor, ROBERT GURALNICK, Department of Mathematics, University of Southern California, Los Angeles, CA 90089-1113; email: guralnic@math.usc.edu.

Titles in This Series

828 **Joel Berman and Paweł M. Idziak,** Generative complexity in algebra, 2005
827 **Trevor A. Welsh,** Fermionic expressions for minimal model Virasoro characters, 2005
826 **Guy Métivier and Kevin Zumbrun,** Large viscous boundary layers for noncharacteristic nonlinear hyperbolic problems, 2005
825 **Yaozhong Hu,** Integral transformations and anticipative calculus for fractional Brownian motions, 2005
824 **Luen-Chau Li and Serge Parmentier,** On dynamical Poisson groupoids I, 2005
823 **Claus Mokler,** An analogue of a reductive algebraic monoid whose unit group is a Kac-Moody group, 2005
822 **Stefano Pigola, Marco Rigoli, and Alberto G. Setti,** Maximum principles on Riemannian manifolds and applications, 2005
821 **Nicole Bopp and Hubert Rubenthaler,** Local zeta functions attached to the minimal spherical series for a class of symmetric spaces, 2005
820 **Vadim A. Kaimanovich and Mikhail Lyubich,** Conformal and harmonic measures on laminations associated with rational maps, 2005
819 **F. Andreatta and E. Z. Goren,** Hilbert modular forms: Mod p and p-adic aspects, 2005
818 **Tom De Medts,** An algebraic structure for Moufang quadrangles, 2005
817 **Javier Fernández de Bobadilla,** Moduli spaces of polynomials in two variables, 2005
816 **Francis Clarke,** Necessary conditions in dynamic optimization, 2005
815 **Martin Bendersky and Donald M. Davis,** V_1-periodic homotopy groups of $SO(n)$, 2004
814 **Johannes Huebschmann,** Kähler spaces, nilpotent orbits, and singular reduction, 2004
813 **Jeff Groah and Blake Temple,** Shock-wave solutions of the Einstein equations with perfect fluid sources: Existence and consistency by a locally inertial Glimm scheme, 2004
812 **Richard D. Canary and Darryl McCullough,** Homotopy equivalences of 3-manifolds and deformation theory of Kleinian groups, 2004
811 **Ottmar Loos and Erhard Neher,** Locally finite root systems, 2004
810 **W. N. Everitt and L. Markus,** Infinite dimensional complex symplectic spaces, 2004
809 **J. T. Cox, D. A. Dawson, and A. Greven,** Mutually catalytic super branching random walks: Large finite systems and renormalization analysis, 2004
808 **Hagen Meltzer,** Exceptional vector bundles, tilting sheaves and tilting complexes for weighted projective lines, 2004
807 **Carlos A. Cabrelli, Christopher Heil, and Ursula M. Molter,** Self-similarity and multiwavelets in higher dimensions, 2004
806 **Spiros A. Argyros and Andreas Tolias,** Methods in the theory of hereditarily indecomposable Banach spaces, 2004
805 **Philip L. Bowers and Kenneth Stephenson,** Uniformizing dessins and Belyĭ maps via circle packing, 2004
804 **A. Yu Ol'shanskii and M. V. Sapir,** The conjugacy problem and Higman embeddings, 2004
803 **Michael Field and Matthew Nicol,** Ergodic theory of equivariant diffeomorphisms: Markov partitions and stable ergodicity, 2004
802 **Martin W. Liebeck and Gary M. Seitz,** The maximal subgroups of positive dimension in exceptional algebraic groups, 2004
801 **Fabio Ancona and Andrea Marson,** Well-posedness for general 2×2 systems of conservation law, 2004
800 **V. Poénaru and C. Tanas,** Equivariant, almost-arborescent representation of open simply-connected 3-manifolds; A finiteness result, 2004
799 **Barry Mazur and Karl Rubin,** Kolyvagin systems, 2004